Math and Test Taking
Grade 6

A Best Value Book™

Written by
Dawn Talluto Jacobi

Edited by
Aaron Levy

© Carson-Dellosa CD-3756

ISBN 0-88724-537-4

Table of Contents

Number Concepts

Number Words...1
Mathematical Expressions...............................2
Mathematical Expressions...............................3
Mathematical Expressions...............................4
Mathematical Expressions...............................5
Mathematical Expressions...............................6
Place Value...7
Place Value...8
Place Value...9
Place Value...10
Rounding..11
Rounding..12
Fractions (Least Common Denominator).........13
Fractions (Comparison)................................14
Fractions (Simplifying).................................15
Fractions (Equivalents)................................16
Prime Numbers...17
Prime Factorization.....................................18
Variables..19
Number Concepts Practice....................20-27

Computation

Addition/Subtraction of Whole Numbers........28
Multiplication of Whole Numbers...................29
Division of Whole Numbers...........................30
Estimation (Whole Numbers)........................31
Addition of Decimals....................................32
Subtraction of Decimals...............................33
Multiplication of Decimals.............................34
Division of Decimals.....................................35
Decimals Mixed Practice..............................36
Estimation (Decimals)..................................37
Addition of Fractions and Mixed Numbers......38
Subtraction of Fractions...............................39
Subtraction of Mixed Numbers......................40
Multiplication of Fractions.............................41
Multiplication of Mixed Numbers...................42
Division of Fractions....................................43
Division of Mixed Numbers...........................44
Fraction to Decimal.....................................45
Decimal to Fraction.....................................46
Decimal to Percent......................................47
Percent to Decimal......................................48
Fraction to Percent......................................49
Percent to Fraction......................................50
Fractions/Decimals/Percentages..................51
Estimation by Rounding...............................52
Estimation by Rounding...............................53
Percentages...54
Percentages...55
Percentages...56
Percentages...57
Equations...58
Equations...59

Equations...60
Equations...61
Substitution..62
Substitution..63
Proportions..64
Proportions..65
Computation Practice.............................66-73

Application

Word Problems..74
Coordinate Points..75
Coordinate Points..76
Money..77
Money..78
Averages..79
Proportions..80
Probability..81
Number Sentences......................................82
Number Sentences......................................83
Geometry (Angles)......................................84
Geometry (Angles)......................................85
Geometry (Triangles)...................................86
Geometry (Circles)......................................87
Perimeter..88
Area...89
Volume...90
Frequency Table..91
Frequency Table..92
Fraction Word Problems...............................93
Fraction Word Problems...............................94
Percentage Word Problems..........................95
Time Word Problems....................................96
Metric Equivalents.......................................97
Bar Graph...98
Line Graph..99
Mean, Median, Mode, Range.......................100
Application Practice....................................101
Application Practice....................................102
Application Practice....................................103
Application Practice....................................104
Application Practice....................................105
Application Practice....................................106
Application Practice....................................107
Application Practice....................................108

Practice Tests

Practice Test...109
Practice Test...110
Practice Test...111
Practice Test...112
Practice Test...113
Practice Test...114
Practice Test...115
Practice Test...116
Answer Key...117-122
Student Answer Sheet...........................123-124

Math and Test Taking

Written by Dawn Talluto Jacobi
Edited by Aaron Levy

About the Author

Dawn Talluto Jacobi is currently working as a math teacher and teacher of the gifted at Destrehan high School in Destrehan, Louisiana. She earned a degree in Mathematics from the University of New Orleans and is currently pursuing a Master's degree in Gifted Education. Eternally grateful for the many blessings bestowed upon her, Dawn thanks her husband, Jimmy, and their four beautiful children, Kara, Eric, Kaitlin, and Matthew, for their inspiration and support.

Perfect for school or home, every **Kelley Wingate Best Value Book**™ has been designed to help students master the skills necessary to succeed. Each book is packed with reproducible test pages, 96 cut-apart flash cards, and supplemental resource pages full of valuable information, ideas, and activities. These activities may be used as classroom or homework activities, or as enrichment material for a math program.

The purpose of this book is to provide conceptual, computational, and applied mathematical skills practice while reinforcing positive test-taking strategies. The format and types of activities have been patterned after those in standardized tests such as Stanford Achievement, LEAP, Iowa Test of Basic Skills, and other state and national achievement tests. The activities have been sequenced to facilitate successful completion of the assigned tasks, thus building the confidence and self-esteem students need to meet academic challenges.

The practice tests in this book cover the range of cognitive skills from basic concepts, to computation skills, to applied mathematical concepts. Practice tests at the end of each section provide opportunities for cumulative review. These practice tests may be administered in various ways. One method is to give students the activities consecutively, each cognitive skill building upon the previous. Alternately, select one sheet from each skill category and give students this packet, a comprehensive approach similar to many standardized tests. Once students are familiar with the practice test format, consider giving them timed practice tests, since many standardized tests are timed. Take into consideration the length and difficulty of the test, as well as the competency level of the test-takers. Cumulative practice tests are included at the end of each skill section, and at the end of the book. Extra tests may be assembled by selecting one or two pages from each of the three skill areas and administering them together.

Flash Card Ideas and Activities

   Included in the back of this book are 96 flash cards ideal for individual review, group solving sessions, or as part of timed, sequential, or grouped tests. Pull out the flash cards and cut them apart or, if you have access to a paper cutter, use that to cut them into individual cards. Here are just a few of the ways you may want to use these flash cards:

- Play "Around the World with Flash Cards," the object of which is to be the first student to circle the room and return to his own seat. Have two students stand at their desks. Show them one flash card. The first to correctly answer the flash card problem advances and stands beside the next seat in line. The student in that seat stands. Show these two students a flash card, and repeat the process. The winner continues to advance and challenge, in which case he sits down and the new winner advances to face a new flash card and the next student.

- Hold a "Math Challenge." Divide the class into two teams and have a representative from each team stand at the front of the room beside a desk and attempt to be the first to answer a flash card problem. The student who is able to answer the question first can tap the desk (or ring a bell) to signal readiness to give an answer. Award points for correct answers. If the student answers incorrectly, allow the other team a chance to answer.

- Use a timer or stopwatch to record how many problems a student can answer correctly in a certain amount of time. Review incorrect answers and repeat the exercise. Provide rewards for improved scores.

- Give students a card with a math fact on it. Have them write out the other members of that fact family, or brainstorm other math problems that have the same answer.

- Use flash cards as impromptu quizzes. Give each student three to five cards attached to an answer sheet that he can complete and return. Vary the selection of cards given to each student for each quiz.

- Post a certain number of cards, daily or weekly, as bonus questions or for extra credit.

Name _____

Directions

Read each question and choose the correct answer. Mark the space for the answer you have chosen. Mark NH if the answer is not here.

1. What is the numeral for seventy thousand, fourteen?

   a. 7,014
   b. 71,014
   c. 70,104
   d. 70,014
   e. NH

6. What is the numeral for thirteen thousand, eleven?

   f. 1,311
   g. 13,011
   h. 130,011
   j. 1,300,011
   k. NH

2. What is the numeral for eight thousand, three hundred forty?

   f. 8,340
   g. 80,340
   h. 803,040
   j. 830,040
   k. NH

7. What is the numeral for four thousand, two hundred five?

   a. 425
   b. 4,025
   c. 4,205
   d. 4,250
   e. NH

3. What is the numeral for fifteen thousand, seven hundred four?

   a. 1,574
   b. 15,704
   c. 15,740
   d. 150,704
   e. NH

8. What is the numeral for nine thousand, three hundred seventy-eight?

   f. 90,378
   g. 900,378
   h. 93,078
   j. 9,378
   k. NH

4. What is the numeral for twelve thousand, two?

   f. 1,202
   g. 12,020
   h. 12,002
   j. 120,002
   k. NH

9. What is the numeral for one hundred thousand, forty?

   a. 100,040
   b. 10,400
   c. 10,040
   d. 1,040
   e. NH

5. What is the numeral for one thousand, four hundred sixty-one?

   a. 130,061
   b. 14,061
   c. 10,461
   d. 1,461
   e. NH

10. What is the numeral for twenty-three thousand, sixty-eight?

   f. 2,368
   g. 23,068
   h. 23,086
   j. 230,068
   k. NH

Directions
Read each question and choose the correct answer. Mark the space for the answer you have chosen. Mark NH if the answer is not here.

1. What is another way of writing (4 + 5) + 2?

   a. 4 + (4 + 2)
   b. 2 + (4 + 5)
   c. (5 + 4) + (2 + 5)
   d. (4 + 2) + (5 + 2)
   e. NH

6. What is another way of writing 5,281?

   f. 5 x 2 + 8 x 1
   g. 500 + 20 + 81
   h. 5,000 + 200 + 80 + 1
   j. (5 + 1,000) + (2 + 100) + (8 + 10) + 1
   k. NH

2. What is another way of writing 57?

   f. 50 + 7
   g. 5 + 7
   h. 5 x 7
   j. 5 + 2 x 7
   k. NH

7. What is another way of writing
   (3 x 2) + (4 x 8)?

   a. 5 + 12
   b. 38
   c. (3 x 4) + (2 x 8)
   d. 17
   e. NH

3. What is another way of writing 351?

   a. 3 x 5 x 1
   b. 3 + 100 x 51
   c. 3 x 5 + 3 x 1
   d. 300 + 50 + 1
   e. NH

8. What is another way of writing (7 + 3) + 2?

   f. 10 x 2
   g. 7 x 3 x 2
   h. 7 + (3 + 2)
   j. 20
   k. NH

4. What is another way of writing
   (5 + 3) x (4 + 5)?

   f. 5 x 4 x 3
   g. 8 x 9
   h. 15 + 20
   j. 20 + 15 + 20 + 12
   k. NH

9. What is another way of writing 9,084?

   a. 9,000 x 84
   b. 90 x 84
   c. 9 + 0 + 8 + 4
   d. 9,000 + 80 + 4
   e. NH

5. What is another way of writing 3 x (2 + 4)?

   a. 6 + 4
   b. (3 x 2) + 4
   c. (3 x 2) + (3 x 4)
   d. 36
   e. NH

10. What is another way of writing 17,608?

   f. 10,000 + 7,000 + 600 + 8
   g. 17 x 608
   h. 1 + 7 + 6 + 0 + 8
   j. 1,000 x 176
   k. NH

Directions
Read each question and choose the correct answer. Mark the space for the answer you have chosen. Mark NH if the answer is not here.

1. What is another name for $\frac{11}{2}$?

   a. $5\frac{1}{2}$ d. $6\frac{1}{2}$

   b. $1\frac{1}{2}$ e. NH

   c. $\frac{2}{11}$

6. What is another name for $9\frac{7}{8}$?

   f. $\frac{24}{8}$ j. $\frac{63}{8}$

   g. $\frac{79}{7}$ k. NH

   h. $\frac{79}{8}$

2. What is another name for $\frac{25}{4}$?

   f. $3\frac{1}{4}$ j. $2\frac{5}{4}$

   g. $5\frac{1}{2}$ k. NH

   h. $6\frac{1}{4}$

7. What is another name for $7\frac{2}{3}$?

   a. $\frac{23}{2}$ d. $\frac{72}{3}$

   b. $8\frac{1}{2}$ e. NH

   c. $6\frac{4}{3}$

3. What is another name for $\frac{100}{3}$?

   a. $100\frac{1}{3}$ d. 3

   b. $33\frac{1}{3}$ e. NH

   c. $3\frac{1}{3}$

8. What fraction is another name for $5\frac{1}{8}$?

   f. $\frac{51}{8}$ j. $\frac{14}{8}$

   g. $\frac{41}{8}$ k. NH

   h. $\frac{32}{5}$

4. What is another name for $\frac{43}{7}$?

   f. $4\frac{3}{7}$ j. $7\frac{1}{6}$

   g. $6\frac{1}{7}$ k. NH

   h. $6\frac{6}{7}$

9. What fraction is another name for $12\frac{3}{4}$?

   a. $\frac{52}{4}$ d. $\frac{42}{3}$

   b. $\frac{51}{4}$ e. NH

   c. $\frac{19}{4}$

5. What is another name for $\frac{98}{5}$?

   a. $9\frac{1}{2}$ d. $19\frac{2}{4}$

   b. $5\frac{1}{4}$ e. NH

   c. $18\frac{2}{3}$

10. What fraction is another name for $22\frac{1}{3}$?

   f. $\frac{25}{3}$ j. $\frac{66}{3}$

   g. $\frac{67}{3}$ k. NH

   h. $\frac{26}{3}$

Directions
Read each question and choose the correct answer. Mark the space for the answer you have chosen. Mark NH if the answer is not here.

---

1. What is another name for 0.7?

   a. 7   **d.** $\frac{7}{100}$

   b. $\frac{7}{10}$ **e.** NH

   c. 70

---

6. What decimal is another name for $\frac{3}{5}$?

   f. 3.5
   g. 0.35
   h. 0.06
   j. 0.6
   k. NH

---

2. What is another name for 1.25?

   f. 125   **j.** $\frac{1}{25}$

   g. $1\frac{1}{4}$ **k.** NH

   h. $12\frac{1}{2}$

---

7. What decimal is another name for $\frac{3}{4}$?

   a. 0.75
   b. 3.4
   c. 4.3
   d. 0.25
   e. NH

---

3. What is another name for 0.29?

   a. $\frac{29}{100}$ **d.** $\frac{0}{29}$

   b. $\frac{2}{9}$ **e.** NH

   c. $\frac{9}{2}$

---

8. What decimal is another name for $\frac{7}{2}$?

   f. 7.2
   g. 3.5
   h. 2.7
   j. 0.72
   k. NH

---

4. What is another name for 0.009?

   f. $\frac{100}{9}$ **j.** $\frac{9}{1000}$

   g. $\frac{9}{10}$ **k.** NH

   h. $\frac{9}{100}$

---

9. What decimal is another name for $\frac{1}{3}$?

   a. 3
   b. 0.13
   c. 0.0003
   d. $0.33\overline{3}$
   e. NH

---

5. What is another name for 3.2?

   a. $\frac{3}{2}$ **d.** $2\frac{3}{10}$

   b. $\frac{2}{3}$ **e.** NH

   c. $3\frac{1}{5}$

---

10. What decimal is another name for $\frac{9}{10}$?

   f. 9.10
   g. 9.1
   h. 0.9
   j. 0.09
   k. NH

Directions
Read each question and choose the correct answer. Mark the space for the answer you have
chosen. Mark NH if the answer is not here.

---

1. What is another name for $\frac{3}{4}$?

   a. 3.4   d. $\frac{9}{16}$

   b. $\frac{6}{8}$ e. NH

   c. $\frac{1}{2}$

6. What is another name for $\frac{14}{3}$?

   f. $1\frac{4}{3}$ j. 14.3

   g. $4\frac{2}{3}$ k. NH

   h. $2\frac{1}{3}$

---

2. What is another name for 0.08?

   f. $\frac{2}{15}$ j. $\frac{8}{5}$

   g. $\frac{8}{10}$ k. NH

   h. 8.0

7. What is another name for $\frac{28}{35}$?

   a. $\frac{5}{8}$ d. $\frac{2}{3}$

   b. $\frac{3}{4}$ e. NH

   c. $\frac{4}{5}$

---

3. What is another name for $3\frac{1}{4}$?

   a. 3.14   d. 31.4

   b. $\frac{13}{4}$ e. NH

   c. $1\frac{3}{4}$

8. What is another name for $5\frac{3}{4}$?

   f. $5\frac{9}{16}$ j. 5.75

   g. $5\frac{6}{9}$ k. NH

   h. 5.25

---

4. What is another name for 0.5?

   f. $\frac{1}{2}$ j. $\frac{1}{4}$

   g. $\frac{5}{100}$ k. NH

   h. $\frac{1}{3}$

9. What is another name for $\frac{72}{80}$?

   a. $\frac{3}{4}$ d. $\frac{9}{10}$

   b. $\frac{4}{5}$ e. NH

   c. $\frac{7}{8}$

---

5. What is another name for $\frac{4}{5}$?

   a. $\frac{16}{25}$ d. 0.45

   b. 0.60   e. NH

   c. $\frac{16}{20}$

10. What is another name for 0.6?

   f. $\frac{6}{100}$ j. $\frac{4}{5}$

   g. $\frac{3}{5}$ k. NH

   h. $\frac{8}{9}$

Directions
Read each question and choose the correct answer. Mark the space for the answer you have chosen. Mark NH if the answer is not here.

---

1. What fraction is another name for $\frac{5}{6}$?

    a. $\frac{40}{48}$ **d.** $\frac{15}{36}$

    b. $\frac{10}{16}$ **e.** NH

    c. $\frac{8}{11}$

6. What fraction is another name for $\frac{24}{60}$?

    f. $\frac{2}{3}$ **j.** $\frac{3}{5}$

    g. $\frac{4}{5}$ **k.** NH

    h. $\frac{2}{5}$

---

2. What fraction is another name for $\frac{130}{260}$?

    f. $\frac{1}{3}$ **j.** $\frac{1}{5}$

    g. $\frac{1}{2}$ **k.** NH

    h. $\frac{1}{4}$

7. What fraction is another name for $8\frac{3}{4}$?

    a. $\frac{35}{4}$ **d.** $\frac{39}{4}$

    b. $\frac{83}{4}$ **e.** NH

    c. $\frac{31}{4}$

---

3. What fraction is another name for $1\frac{32}{44}$?

    a. $\frac{1}{3}$ **d.** $\frac{1}{5}$

    b. $\frac{1}{2}$ **e.** NH

    c. $\frac{1}{4}$

8. What fraction is another name for $5\frac{3}{15}$?

    f. $\frac{75}{3}$ **j.** $\frac{39}{4}$

    g. $\frac{78}{5}$ **k.** NH

    h. $\frac{78}{15}$

---

4. What fraction is another name for $\frac{48}{72}$?

    f. $\frac{1}{3}$ **j.** $\frac{1}{5}$

    g. $\frac{2}{3}$ **k.** NH

    h. $\frac{1}{4}$

9. What fraction is another name for $\frac{43}{6}$?

    a. $5\frac{1}{2}$ **d.** $7\frac{1}{6}$

    b. $6\frac{5}{6}$ **e.** NH

    c. $7\frac{1}{2}$

---

5. What fraction is another name for $\frac{28}{3}$?

    a. $\frac{1}{3}$ **d.** $\frac{1}{5}$

    b. $\frac{1}{2}$ **e.** NH

    c. $\frac{1}{4}$

10. What fraction is another name for $\frac{42}{54}$?

    f. $\frac{9}{7}$ **j.** $\frac{6}{9}$

    g. $\frac{7}{9}$ **k.** NH

    h. $\frac{7}{8}$

Name _____

Directions
Read each question and choose the correct answer. Mark the space for the answer you have
chosen. Mark NH if the answer is not here.

1. What does the 6 in 86,523 represent?

    a. 60,000
    b. 6,000
    c. 600
    d. 60
    e. NH

2. What does the 4 in 49,751 represent?

    f. 40,000
    g. 4,000
    h. 400
    j. 40
    k. NH

3. What does the 9 in 5,196 represent?

    a. 9,000
    b. 900
    c. 90
    d. 9
    e. NH

4. What does the 1 in 143,206 represent?

    f. 100
    g. 1,000
    h. 10,000
    j. 100,000
    k. NH

5. What does the 3 in 167,534 represent?

    a. 3
    b. 30
    c. 300
    d. 3000
    e. NH

6. In the number 50,316, what digit is in the
    tens place?

    f. 6
    g. 1
    h. 3
    j. 5
    k. NH

7. In the number 412,367, what digit is in the
    hundreds place?

    a. 4
    b. 1
    c. 2
    d. 3
    e. NH

8. In the number 847,695, what digit is in the
    ten thousands place?

    f. 8
    g. 4
    h. 7
    j. 6
    k. NH

9. In the number 90,354, what digit is in the
    ones place?

    a. 0
    b. 3
    c. 5
    d. 4
    e. NH

10. In the number 327,518, what digit is in the
    thousands place?

    f. 2
    g. 7
    h. 3
    j. 5
    k. NH

Name _____

Directions

Read each question and choose the correct answer. Mark the space for the answer you have chosen. Mark NH if the answer is not here.

---

1. What does the 6 in 5.64 represent?

    a. 6 thousandths
    b. 6 hundredths
    c. 6 tenths
    d. 6 ones
    e. NH

6. What number is 2 tenths more than 25.73?

    f. 27.73
    g. 25.75
    h. 25.93
    j. 27.95
    k. NH

---

2. What does the 2 in 16.523 represent?

    f. 2 hundredths
    g. 2 tenths
    h. 2 ones
    j. 2 tens
    k. NH

7. What number is 1 tenth more than 18.6?

    a. 18.7
    b. 28.6
    c. 19.6
    d. 28.7
    e. NH

---

3. What does the 2 in 35.502 represent?

    a. 2 ten thousandths
    b. 2 thousandths
    c. 2 hundredths
    d. 2 tenths
    e. NH

8. What number is 5 tenths more than 526.148?

    f. 576.148
    g. 531.148
    h. 526.198
    j. 526.648
    k. NH

---

4. What does the 8 in 502.3849 represent?

    f. 8 ten thousandths
    g. 8 thousandths
    h. 8 hundredths
    j. 8 tenths
    k. NH

9. What number is 1 hundredth more than 5.324?

    a. 5.424
    b. 5.334
    c. 5.325
    d. 6.324
    e. NH

---

5. What does the 5 in 0.543 represent?

    a. 5 ten thousandths
    b. 5 thousandths
    c. 5 hundredths
    d. 5 tenths
    e. NH

10. What number is 3 hundredths more than 23.224?

    f. 23.524
    g. 23.254
    h. 23.227
    j. 23.557
    k. NH

Name _____

Directions
Read each question and choose the correct answer. Mark the space for the answer you have chosen. Mark NH if the answer is not here.

1. What is the place value of the 3 in 73,412?

   a. ten thousands
   b. thousands
   c. hundreds
   d. tens
   e. NH

6. What number has the 7 in the tenths place?

   f. 572.3
   g. 318.74
   h. 927.21
   j. 123.47
   k. NH

2. What is the place value of the 0 in 42,032?

   f. tens
   g. hundreds
   h. thousands
   j. ten thousands
   k. NH

7. What number has the 4 in the hundreds place?

   a. 5,432.18
   b. 704.8
   c. 49.81
   d. 918.345
   e. NH

3. What is the place value of the 8 in 5.836?

   a. tenths
   b. ones
   c. tens
   d. hundreds
   e. NH

8. What number has the 3 in the ones place?

   f. 125.32
   g. 765.03
   h. 304.5
   j. 593.8
   k. NH

4. What is the place value of the 5 in 325.81?

   f. hundreds
   g. tens
   h. ones
   j. tenths
   k. NH

9. What number has the 2 in the hundredths place?

   a. 1,297.5
   b. 84.127
   c. 543.29
   d. 215.8
   e. NH

5. What is the place value of the 1 in 415,387.9?

   a. tens
   b. hundreds
   c. thousands
   d. ten thousands
   e. NH

10. What number has the 6 in the ten thousands place?

   f. 4,653.2
   g. 61,847.5
   h. 196,312.7
   j. 8,615.4
   k. NH

Directions

Read each question and choose the correct answer. Mark the space for the answer you have chosen. Mark NH if the answer is not here.

---

1. Which number has the greatest value?

   a. 0.04
   b. 0.47
   c. 0.047
   d. 0.074
   e. NH

6. Which number has the least value?

   f. 10.90
   g. 19.00
   h. 10.09
   j. 10.99
   k. NH

---

2. Which number has the greatest value?

   f. 0.134
   g. 0.014
   h. 0.041
   j. 0.043
   k. NH

7. Which number has the least value?

   a. 4.141
   b. 4.411
   c. 4.114
   d. 4.414
   e. NH

---

3. Which number has the greatest value?

   a. 1.001
   b. 1.101
   c. 1.011
   d. 1.100
   e. NH

8. Which number has the least value?

   f. 7.700
   g. 7.070
   h. 7.077
   j. 7.007
   k. NH

---

4. Which number has the greatest value?

   f. 7.844
   g. 7.448
   h. 7.488
   j. 7.848
   k. NH

9. Which number has the least value?

   a. 1.040
   b. 1.400
   c. 1.004
   d. 1.044
   e. NH

---

5. Which number has the greatest value?

   a. 1.08
   b. 1.48
   c. 1.84
   d. 1.80
   e. NH

10. Which number has the least value?

   f. 3.0012
   g. 3.0102
   h. 2.0021
   j. 2.0201
   k. NH

Name _____   Skill: Rounding

Directions
Read each question and choose the correct answer. Mark the space for the answer you have chosen. Mark NH if the answer is not here.

---

1. What is 352 rounded to the nearest ten?

  a. 400
  b. 300
  c. 360
  d. 350
  e. NH

6. What is 1,868 rounded to the nearest ten?

  f. 1,870
  g. 1,860
  h. 1,900
  j. 2,000
  k. NH

---

2. What is 51,836 rounded to the nearest thousand?

  f. 51,800
  g. 52,000
  h. 51,000
  j. 50,000
  k. NH

7. What is 1,351 rounded to the nearest hundred?

  a. 1,300
  b. 1,350
  c. 1,400
  d. 1,360
  e. NH

---

3. What is 586 rounded to the nearest hundred?

  a. 600
  b. 500
  c. 590
  d. 580
  e. NH

8. What is 9,736 rounded to the nearest ten?

  f. 10,000
  g. 9,800
  h. 9,740
  j. 9,730
  k. NH

---

4. What is 1,246 rounded to the nearest hundred?

  f. 1,300
  g. 1,200
  h. 1,250
  j. 1,240
  k. NH

9. What is 4,087 rounded to the nearest hundred?

  a. 4,000
  b. 4,080
  c. 4,090
  d. 4,100
  e. NH

---

5. What is 47,389 rounded to the nearest thousand?

  a. 47,400
  b. 47,390
  c. 47,000
  d. 48,000
  e. NH

10. What is 615.3 rounded to the nearest ten?

  f. 610
  g. 620
  h. 615
  j. 600
  k. NH

Name _____

Directions

Read each question and choose the correct answer. Mark the space for the answer you have chosen. Mark NH if the answer is not here.

1. What is 72.4 rounded to the nearest whole number?

   a. 70
   b. 72
   c. 73
   d. 74
   e. NH

6. What is 37.5 rounded to the nearest whole number?

   f. 40
   g. 39
   h. 38
   j. 37
   k. NH

2. What is 15.39 rounded to the nearest whole number?

   f. 20
   g. 19
   h. 16
   j. 15
   k. NH

7. What is 8.19 rounded to the nearest whole number?

   a. 9
   b. 8
   c. 7
   d. 6
   e. NH

3. What is 26.08 rounded to the nearest whole number?

   a. 28
   b. 27
   c. 26
   d. 25
   e. NH

8. What is 43.81 rounded to the nearest whole number?

   f. 44
   g. 43
   h. 42
   j. 41
   k. NH

4. What is 4.7 rounded to the nearest whole number?

   f. 7
   g. 6
   h. 5
   j. 4
   k. NH

9. What is 610.5 rounded to the nearest whole number?

   a. 609
   b. 610
   c. 611
   d. 612
   e. NH

5. What is 890.6 rounded to the nearest whole number?

   a. 891
   b. 890
   c. 889
   d. 888
   e. NH

10. What is 111.1 rounded to the nearest whole number?

   f. 111
   g. 110
   h. 101
   j. 100
   k. NH

Name _____

Directions

Read each question and choose the correct answer. Mark the space for the answer you have chosen. Mark NH if the answer is not here.

1. What is the least common denominator for $\frac{1}{4}$ and $\frac{5}{6}$?

   a. 24
   b. 18
   c. 12
   d. 8
   e. NH

6. What is the least common denominator for $\frac{3}{5}$ and $\frac{5}{6}$?

   f. 30
   g. 60
   h. 90
   j. 115
   k. NH

2. What is the least common denominator for $\frac{2}{3}$ and $\frac{1}{8}$?

   f. 12
   g. 16
   h. 18
   j. 24
   k. NH

7. What is the least common denominator for $\frac{7}{8}$ and $\frac{5}{12}$?

   a. 12
   b. 24
   c. 48
   d. 96
   e. NH

3. What is the least common denominator for $\frac{1}{6}$ and $\frac{7}{8}$?

   a. 12
   b. 16
   c. 48
   d. 24
   e. NH

8. What is the least common denominator for $\frac{3}{4}$ and $\frac{5}{9}$?

   f. 36
   g. 72
   h. 81
   j. 144
   k. NH

4. What is the least common denominator for $\frac{1}{2}$ and $\frac{2}{5}$?

   f. 5
   g. 10
   h. 15
   j. 20
   k. NH

9. What is the least common denominator for $\frac{5}{6}$ and $\frac{1}{9}$?

   a. 9
   b. 16
   c. 18
   d. 54
   e. NH

5. What is the least common denominator for $\frac{3}{4}$ and $\frac{1}{3}$?

   a. 12
   b. 15
   c. 16
   d. 24
   e. NH

10. What is the least common denominator for $\frac{2}{3}$ and $\frac{4}{5}$?

   f. 9
   g. 12
   h. 15
   j. 30
   k. NH

Name _____

Directions
Read each question and choose the correct answer. Mark the space for the answer you have chosen. Mark NH if the answer is not here.

1. Which fraction is smallest?

   a. $\frac{2}{3}$ d. $\frac{1}{2}$

   b. $\frac{1}{3}$ e. NH

   c. $\frac{3}{4}$

6. Which fraction is greatest?

   f. $\frac{5}{6}$ j. $\frac{1}{3}$

   g. $\frac{3}{7}$ k. NH

   h. $\frac{4}{9}$

2. Which fraction is smallest?

   f. $\frac{2}{5}$ j. $\frac{2}{3}$

   g. $\frac{4}{5}$ k. NH

   h. $\frac{5}{6}$

7. Which fraction is greatest?

   a. $\frac{1}{2}$ d. $\frac{4}{10}$

   b. $\frac{7}{9}$ e. NH

   c. $\frac{3}{6}$

3. Which fraction is smallest?

   a. $\frac{3}{10}$ d. $\frac{4}{5}$

   b. $\frac{1}{5}$ e. NH

   c. $\frac{10}{20}$

8. Which fraction is greatest?

   f. $\frac{4}{10}$ j. $\frac{3}{4}$

   g. $\frac{9}{18}$ k. NH

   h. $\frac{8}{10}$

4. Which fraction is smallest?

   f. $\frac{7}{10}$ j. $\frac{5}{6}$

   g. $\frac{4}{9}$ k. NH

   h. $\frac{6}{8}$

9. Which fraction is greatest?

   a. $\frac{12}{15}$ d. $\frac{7}{8}$

   b. $\frac{7}{14}$ e. NH

   c. $\frac{3}{4}$

5. Which fraction is smallest?

   a. $\frac{4}{8}$ d. $\frac{1}{2}$

   b. $\frac{3}{5}$ e. NH

   c. $\frac{1}{3}$

10. Which fraction is greatest?

   f. $\frac{9}{10}$ j. $\frac{50}{100}$

   g. $\frac{1}{50}$ k. NH

   h. $\frac{3}{6}$

Directions

Read each question and choose the correct answer. Mark the space for the answer you have chosen. Mark NH if the answer is not here.

1. Which fraction is in its simplest form?

  a. $\frac{4}{9}$ d. $\frac{9}{15}$

  b. $\frac{6}{9}$ e. NH

  c. $\frac{3}{21}$

6. What is the simplest form of $\frac{48}{60}$?

  f. $\frac{24}{30}$ j. $\frac{8}{10}$

  g. $\frac{12}{15}$ k. NH

  h. $\frac{4}{5}$

2. Which fraction is in its simplest form?

  f. $\frac{9}{12}$ j. $\frac{12}{15}$

  g. $\frac{2}{22}$ k. NH

  h. $\frac{1}{8}$

7. What is the simplest form of $\frac{24}{32}$?

  a. $\frac{6}{8}$ d. $\frac{12}{16}$

  b. $\frac{12}{18}$ e. NH

  c. $\frac{3}{4}$

3. Which fraction is in its simplest form?

  a. $\frac{3}{15}$ d. $\frac{2}{10}$

  b. $\frac{3}{5}$ e. NH

  c. $\frac{18}{21}$

8. What is the simplest form of $\frac{40}{80}$?

  f. $\frac{4}{8}$ j. $\frac{1}{2}$

  g. $\frac{20}{40}$ k. NH

  h. $\frac{10}{20}$

4. Which fraction is in its simplest form?

  f. $\frac{16}{18}$ j. $\frac{18}{27}$

  g. $\frac{4}{5}$ k. NH

  h. $\frac{15}{20}$

9. What is the simplest form of $\frac{54}{72}$?

  a. $\frac{9}{12}$ d. $\frac{3}{4}$

  b. $\frac{18}{24}$ e. NH

  c. $\frac{27}{30}$

5. Which fraction is in its simplest form?

  a. $\frac{6}{7}$ d. $\frac{6}{18}$

  b. $\frac{9}{15}$ e. NH

  c. $\frac{2}{4}$

10. What is the simplest form of $\frac{36}{144}$?

  f. $\frac{2}{8}$ j. $\frac{3}{12}$

  g. $\frac{1}{4}$ k. NH

  h. $\frac{3}{6}$

Directions
Read each question and choose the correct answer. Mark the space for the answer you have chosen. Mark NH if the answer is not here.

1. Which fraction is <u>not</u> equivalent to $\frac{3}{4}$?

   a. $\frac{6}{8}$ d. $\frac{9}{16}$

   b. $\frac{30}{40}$ e. NH

   c. $\frac{15}{20}$

6. Which fraction is <u>not</u> equivalent to $\frac{3}{8}$?

   f. $\frac{9}{24}$ j. $\frac{21}{56}$

   g. $\frac{28}{64}$ k. NH

   h. $\frac{27}{72}$

2. Which fraction is <u>not</u> equivalent to $\frac{8}{9}$?

   f. $\frac{64}{81}$ j. $\frac{16}{18}$

   g. $\frac{40}{45}$ k. NH

   h. $\frac{96}{108}$

7. Which fraction is <u>not</u> equivalent to $\frac{2}{7}$?

   a. $\frac{14}{49}$ d. $\frac{18}{63}$

   b. $\frac{4}{49}$ e. NH

   c. $\frac{6}{21}$

3. Which fraction is <u>not</u> equivalent to $\frac{5}{8}$?

   a. $\frac{50}{80}$ d. $\frac{45}{72}$

   b. $\frac{25}{64}$ e. NH

   c. $\frac{25}{40}$

8. Which fraction is <u>not</u> equivalent to $\frac{1}{4}$?

   f. $\frac{4}{16}$ j. $\frac{7}{28}$

   g. $\frac{9}{36}$ k. NH

   h. $\frac{4}{44}$

4. Which fraction is <u>not</u> equivalent to $\frac{1}{6}$?

   f. $\frac{4}{24}$ j. $\frac{8}{48}$

   g. $\frac{9}{54}$ k. NH

   h. $\frac{3}{36}$

9. Which fraction is <u>not</u> equivalent to $\frac{5}{9}$?

   a. $\frac{25}{81}$ d. $\frac{10}{18}$

   b. $\frac{20}{36}$ e. NH

   c. $\frac{45}{81}$

5. Which fraction is <u>not</u> equivalent to $\frac{2}{3}$?

   a. $\frac{10}{15}$ d. $\frac{14}{21}$

   b. $\frac{6}{9}$ e. NH

   c. $\frac{16}{18}$

10. Which fraction is <u>not</u> equivalent to $\frac{7}{10}$?

   f. $\frac{49}{100}$ j. $\frac{49}{70}$

   g. $\frac{42}{60}$ k. NH

   h. $\frac{21}{30}$

Name _____

Directions

Read each question and choose the correct answer. Mark the space for the answer you have chosen. Mark NH if the answer is not here.

1. Which of the following is a prime number?

   a. 15
   b. 21
   c. 19
   d. 12
   e. NH

2. Which of the following is a prime number?

   f. 37
   g. 57
   h. 18
   j. 39
   k. NH

3. Which of the following is a prime number?

   a. 23
   b. 21
   c. 42
   d. 111
   e. NH

4. Which of the following is a prime number?

   f. 49
   g. 56
   h. 27
   j. 59
   k. NH

5. Which of the following is a prime number?

   a. 57
   b. 41
   c. 18
   d. 93
   e. NH

6. Which of the following is a prime number?

   f. 52
   g. 33
   h. 61
   j. 81
   k. NH

7. Which of the following is a prime number?

   a. 51
   b. 77
   c. 9
   d. 17
   e. NH

8. Which of the following is a prime number?

   f. 72
   g. 11
   h. 42
   j. 76
   k. NH

9. Which of the following is a prime number?

   a. 57
   b. 29
   c. 78
   d. 49
   e. NH

10. Which of the following is a prime number?

   f. 13
   g. 16
   h. 78
   j. 96
   k. NH

Directions

Read each question and choose the correct answer. Mark the space for the answer you have chosen. Mark NH if the answer is not here.

1. What is the prime factorization of 36?

   a. 6 x 6
   b. 9 x 4
   c. 3 x 12
   d. 2 x 2 x 3 x 3
   e. NH

6. What is the prime factorization of 56?

   f. 2 x 28
   g. 2 x 2 x 2 x 7
   h. 4 x 14
   j. 7 x 8
   k. NH

2. What is the prime factorization of 60?

   f. 2 x 2 x 3 x 5
   g. 4 x 15
   h. 2 x 30
   j. 6 x 10
   k. NH

7. What is the prime factorization of 100?

   a. 10 x 10
   b. 20 x 5
   c. 2 x 2 x 5 x 5
   d. 25 x 4
   e. NH

3. What is the prime factorization of 18?

   a. 3 x 6
   b. 2 x 3 x 3
   c. 18 x 1
   d. 9 x 2
   e. NH

8. What is the prime factorization of 96?

   f. 8 x 12
   g. 4 x 24
   h. 2 x 48
   j. 2 x 2 x 2 x 2 x 2 x 3
   k. NH

4. What is the prime factorization of 48?

   f. 2 x 24
   g. 2 x 2 x 2 x 2 x 3
   h. 6 x 8
   j. 16 x 3
   k. NH

9. What is the prime factorization of 54?

   a. 6 x 9
   b. 2 x 3 x 3 x 3
   c. 3 x 18
   d. 2 x 27
   e. NH

5. What is the prime factorization of 75?

   a. 75 x 1
   b. 3 x 3 x 5
   c. 3 x 5 x 5
   d. 3 x 25
   e. NH

10. What is the prime factorization of 81?

   f. 3 x 3 x 3 x 3
   g. 1 x 81
   h. 3 x 27
   j. 9 x 9
   k. NH

Directions

Read each question and choose the correct answer. Mark the space for the answer you have chosen. Mark NH if the answer is not here.

1. Solve for n:

   $n + 3 = 8$

   a. 3
   b. 4
   c. 5
   d. 6
   e. NH

2. Solve for n:

   $6 + n = 13$

   f. 5
   g. 6
   h. 7
   j. 8
   k. NH

3. Solve for n:

   $12 + n = 21$

   a. 9
   b. 10
   c. 11
   d. 12
   e. NH

4. Solve for n:

   $n + 4 = 10$

   f. 5
   g. 6
   h. 7
   j. 8
   k. NH

5. Solve for n:

   $4 \times n = 20$

   a. 2
   b. 3
   c. 4
   d. 5
   e. NH

6. Solve for n:

   $n \div 3 = 6$

   f. 17
   g. 18
   h. 19
   j. 20
   k. NH

7. Solve for n:

   $n - 8 = 6$

   a. 12
   b. 13
   c. 14
   d. 15
   e. NH

8. Solve for n:

   $6 \times n = 36$

   f. 6
   g. 7
   h. 8
   j. 9
   k. NH

9. Solve for n:

   $n + 9 = 15$

   a. 4
   b. 5
   c. 6
   d. 7
   e. NH

10. Solve for n:

   $n \div 4 = 3$

   f. 1
   g. 4
   h. 8
   j. 12
   k. NH

Directions

Read each question and choose the correct answer. Mark the space for the answer you have chosen. Mark NH if the answer is not here.

1. What is another name for $\frac{36}{42}$?

   a. $\frac{9}{21}$ d. $\frac{23}{15}$

   b. $\frac{6}{7}$ e. NH

   c. $\frac{7}{8}$

2. What does the 6 in 5,063.2 represent?

   f. 6 hundreds
   g. 6 tens
   h. 6 ones
   j. 6 tenths
   k. NH

3. Which is a prime number?

   a. 3
   b. 6
   c. 9
   d. 15
   e. NH

4. What is another way of writing $\frac{29}{5}$?

   f. $2\frac{7}{5}$ j. $5\frac{2}{5}$

   g. $7\frac{2}{5}$ k. NH

   h. $5\frac{1}{2}$

5. What is the least common denominator for $\frac{2}{3}$ and $\frac{3}{7}$?

   a. 7
   b. 14
   c. 21
   d. 42
   e. NH

6. What does the 7 in 346.78 represent?

   f. 7 hundredths
   g. 7 tens
   h. 7 ones
   j. 7 tenths
   k. NH

7. Solve for n:
   $$7 \times n = 28$$

   a. 7
   b. 6
   c. 5
   d. 4
   e. NH

8. What does the 4 in 86.34 represent?

   f. 4 hundredths
   g. 4 tenths
   h. 4 ones
   j. 4 tens
   k. NH

9. What is 3,246 rounded to the nearest ten?

   a. 3,000
   b. 3,300
   c. 3,250
   d. 3,260
   e. NH

10. What is the numeral for five thousand, fifteen?

    f. 5,105
    g. 5,015
    h. 5,115
    j. 5,515
    k. NH

Name _____

Directions

Read each question and choose the correct answer. Mark the space for the answer you have chosen. Mark NH if the answer is not here.

1. What is another name for $\frac{24}{36}$?

   a. $\frac{2}{20}$ d. $\frac{22}{18}$

   b. $\frac{5}{8}$ e. NH

   c. $\frac{2}{3}$

2. What does the 8 in 2,158.2 represent?

   f. 8 hundreds
   g. 8 tens
   h. 8 ones
   j. 8 tenths
   k. NH

3. Which is a prime number?

   a. 5
   b. 9
   c. 6
   d. 12
   e. NH

4. What is another way of writing $\frac{31}{6}$?

   f. $3\frac{7}{6}$ j. $5\frac{1}{6}$

   g. $8\frac{2}{3}$ k. NH

   h. $5\frac{1}{23}$

5. What is the least common denominator for $\frac{3}{5}$ and $\frac{4}{8}$?

   a. 25
   b. 15
   c. 40
   d. 36
   e. NH

6. What does the 2 in 365.21 represent?

   f. 2 hundredths
   g. 2 tens
   h. 2 ones
   j. 2 tenths
   k. NH

7. Solve for n:
   $$4 \times n = 36$$

   a. 8
   b. 9
   c. 3
   d. 2
   e. NH

8. What does the 5 in 98.45 represent?

   f. 5 hundredths
   g. 5 tenths
   h. 5 ones
   j. 5 tens
   k. NH

9. What is 2,561 rounded to the nearest ten?

   a. 2,000
   b. 2,500
   c. 2,570
   d. 2,560
   e. NH

10. What is the numeral for three thousand, thirty-six?

    f. 3,136
    g. 3,036
    h. 3,306
    j. 3,360
    k. NH

Directions

Read each question and choose the correct answer. Mark the space for the answer you have chosen. Mark NH if the answer is not here.

1. What is 73.84 rounded to the nearest whole number?

   a. 70
   b. 73
   c. 74
   d. 76
   e. NH

2. What is the place value of the 4 in the number 56,403?

   f. hundreds
   g. tens
   h. thousands
   j. ten thousands
   k. NH

3. What digit is in the tens place in the number 6,703.52?

   a. 5
   b. 0
   c. 3
   d. 7
   e. NH

4. Which fraction is in its simplest form?

   f. $\frac{6}{15}$ j. $\frac{6}{7}$

   g. $\frac{7}{14}$ k. NH

   h. $\frac{3}{6}$

5. Solve for n:
        3 + n = 9

   a. 5
   b. 6
   c. 7
   d. 8
   e. NH

6. What is another way of writing 0.19?

   f. $1\frac{9}{10}$ j. $\frac{19}{100}$

   g. $\frac{19}{10}$ k. NH

   h. $\frac{19}{1000}$

7. Which fraction is greatest?

   a. $\frac{5}{6}$ d. $\frac{1}{9}$

   b. $\frac{3}{4}$ e. NH

   c. $\frac{2}{3}$

8. What number is three tenths more than 52.36?

   f. 52.66
   g. 82.36
   h. 52.29
   j. 55.36
   k. NH

9. Which of the following is a prime number?

   a. 15
   b. 18
   c. 21
   d. 29
   e. NH

10. What is another way of writing (5 + 3) + 4?

    f. 53 + 4
    g. 5 + (3 + 4)
    h. (5 + 3) + 95 + 4)
    j. (5 + 4) + (5 + 5)
    k. NH

Directions

Read each question and choose the correct answer. Mark the space for the answer you have chosen. Mark NH if the answer is not here.

---

1. What is 65.78 rounded to the nearest whole number?

   a. 65
   b. 66
   c. 67
   d. 65.8
   e. NH

6. What is another way of writing 0.17?

   f. $1\frac{7}{10}$ j. $\frac{17}{100}$

   g. $\frac{17}{10}$ k. NH

   h. $\frac{17}{1000}$

---

2. What is the place value of the 3 in the number 55,326?

   f. hundreds
   g. tens
   h. thousands
   j. ten thousands
   k. NH

7. Which fraction is greatest?

   a. $\frac{1}{2}$ d. $\frac{2}{6}$

   b. $\frac{1}{3}$ e. NH

   c. $\frac{2}{5}$

---

3. What digit is in the tens place in the number 7,825.36?

   a. 3
   b. 8
   c. 4
   d. 2
   e. NH

8. What number is three tenths more than 55.41?

   f. 52.66
   g. 55.44
   h. 58.41
   j. 55.38
   k. NH

---

4. Which fraction is in its simplest form?

   f. $\frac{2}{14}$ j. $\frac{3}{6}$

   g. $\frac{5}{16}$ k. NH

   h. $\frac{4}{8}$

9. Which of the following is a prime number?

   a. 12
   b. 13
   c. 18
   d. 21
   e. NH

---

5. Solve for n:
   $$5 + n = 12$$

   a. 5
   b. 6
   c. 7
   d. 8
   e. NH

10. What is another way of writing (4 + 6) + 2?

   f. 46 + 2
   g. 4 + (6 + 2)
   h. (4 + 6) + 2 + 30)
   j. (4 + 6) + (2 + 1)
   k. NH

Directions

Read each question and choose the correct answer. Mark the space for the answer you have chosen. Mark NH if the answer is not here.

---

1. Which fraction names the smallest number?

   a. $\frac{2}{5}$ d. $\frac{1}{2}$

   b. $\frac{5}{9}$ e. **NH**

   c. $\frac{3}{4}$

2. What is another way of writing
   $(2 + 7) \times (3 + 1)$?

   f.   **27 x 31**
   g.   **27 + 31**
   h.   **9 x 4**
   j.   **6 + 8**
   k.   **NH**

3. Solve for n:
   $$n - 3 = 8$$

   a.   **14**
   b.   **13**
   c.   **12**
   d.   **11**
   e.   **NH**

4. What is another way of writing $3\frac{4}{9}$?

   f. $3\frac{1}{9}$ j. $\frac{32}{9}$

   g. $3\frac{9}{4}$ k.   **NH**

   h. $9\frac{3}{4}$

5. What is 52,312 rounded to the nearest thousand?

   a.   **50,000**
   b.   **52,000**
   c.   **52,300**
   d.   **52,400**
   e.   **NH**

6. What is another name for $\frac{4}{5}$?

   f. $\frac{8}{25}$ j. $\frac{24}{30}$

   g. $\frac{8}{9}$ k.   **NH**

   h. $\frac{16}{25}$

7. What number is 4 hundredths more than 975.351?

   a.   **975.751**
   b.   **975.391**
   c.   **975.355**
   d.   **975.795**
   e.   **NH**

8. Which fraction is in its simplest form?

   f. $\frac{21}{24}$ j. $\frac{18}{30}$

   g. $\frac{2}{4}$ k.   **NH**

   h. $\frac{5}{8}$

9. What digit is in the tenths place in the number 57,346.18?

   a.   **4**
   b.   **6**
   c.   **1**
   d.   **8**
   e.   **NH**

10. What is another name for 0.18?

   f. $1\frac{8}{10}$ j. $1\frac{8}{100}$

   g. $\frac{18}{10}$ k.   **NH**

   h. $\frac{18}{100}$

Directions

Read each question and choose the correct answer. Mark the space for the answer you have chosen. Mark NH if the answer is not here.

---

1. Which fraction is smallest?

   a. $\frac{2}{3}$ d. $\frac{1}{3}$

   b. $\frac{5}{7}$ e. NH

   c. $\frac{3}{4}$

6. What is another name for $\frac{2}{7}$?

   f. $\frac{8}{25}$ j. $\frac{14}{35}$

   g. $\frac{4}{11}$ k. NH

   h. $\frac{12}{42}$

---

2. What is another way of writing $(3 + 6) \times (2 + 4)$?

   f. 36 x 24
   g. 36 + 24
   h. 9 x 6
   j. 9 + 6
   k. NH

7. What number is 4 hundredths more than 365.257?

   a. 765.257
   b. 769.352
   c. 369.254
   d. 366.795
   e. NH

---

3. Solve for n:
   $$n - 5 = 9$$

   a. 14
   b. 12
   c. 10
   d. 13
   e. NH

8. Which fraction is in its simplest form?

   f. $\frac{12}{14}$ j. $\frac{12}{15}$

   g. $\frac{3}{7}$ k. NH

   h. $\frac{6}{8}$

---

4. What is another way of writing $2\frac{3}{5}$?

   f. $2\frac{1}{5}$ j. $\frac{13}{5}$

   g. $2\frac{1}{3}$ k. NH

   h. $8\frac{3}{5}$

9. What digit is in the tenths place in the number 38,425.63?

   a. 4
   b. 2
   c. 5
   d. 6
   e. NH

---

5. What is 55,561 rounded to the nearest thousand?

   a. 55,600
   b. 56,000
   c. 56,560
   d. 55,570
   e. NH

10. What is another name for 0.19?

   f. $1\frac{9}{10}$ j. $1\frac{9}{100}$

   g. $\frac{19}{10}$ k. NH

   h. $\frac{19}{100}$

Name _____

Directions

Read each question and choose the correct answer. Mark the space for the answer you have chosen. Mark NH if the answer is not here.

---

1. What is another way of writing $\frac{2}{4}$?

   a. 0.5
   b. 0.24
   c. 0.4
   d. 2.4
   e. NH

6. Solve for n:
   $$n \div 2 = 8$$

   f. 16
   g. 12
   h. 8
   j. 4
   k. NH

---

2. Which decimal names the greatest number?

   f. 0.671
   g. 0.761
   h. 0.716
   j. 0.617
   k. NH

7. What fraction is <u>not</u> equivalent to $\frac{5}{6}$?

   a. $\frac{10}{12}$ d. $\frac{40}{48}$

   b. $\frac{25}{36}$ e. NH

   c. $\frac{25}{30}$

---

3. What does the 5 in 35,812.97 represent?

   a. 50
   b. 500
   c. 5,000
   d. 50,000
   e. NH

8. What is the place value of the 7 in the number 3,752.19?

   f. tens
   g. hundreds
   h. tenths
   j. hundredths
   k. NH

---

4. What is the least common denominator for $\frac{1}{6}$ and $\frac{7}{15}$?

   f. 30
   g. 45
   h. 50
   j. 60
   k. NH

9. What is the numeral for nineteen thousand, forty-three?

   a. 10,943
   b. 19,043
   c. 19,403
   d. 1,943
   e. NH

---

5. What does the 8 in 32,846.5 represent?

   a. 8,000
   b. 800
   c. 80
   d. 8
   e. NH

10. What is 813.49 rounded to the nearest whole number?

   f. 800
   g. 810
   h. 813
   j. 814
   k. NH

Directions
Read each question and choose the correct answer. Mark the space for the answer you have chosen. Mark NH if the answer is not here.

1. What is another way of writing $\frac{1}{5}$?

   a. 0.5
   b. 0.20
   c. 0.15
   d. 1.5
   e. NH

2. Which decimal names the greatest number?

   f. 0.523
   g. 0.513
   h. 0.534
   j. 0.32
   k. NH

3. What does the 5 in 46,253.78 represent?

   a. 50
   b. 500
   c. 5000
   d. 50000
   e. NH

4. What is the least common denominator for $\frac{1}{3}$ and $\frac{7}{5}$?

   f. 18
   g. 15
   h. 21
   j. 8
   k. NH

5. What does the 4 in 41,563.72 represent?

   a. 40,000
   b. 4,000
   c. 40
   d. 4
   e. NH

6. Solve for n:
   $$n \div 3 = 9$$

   f. 15
   g. 27
   h. 17
   j. 6
   k. NH

7. Which fraction is <u>not</u> equivalent to $\frac{3}{5}$?

   a. $\frac{6}{10}$ d. $\frac{15}{30}$

   b. $\frac{9}{15}$ e. NH

   c. $\frac{12}{20}$

8. What is the place value of the 8 in the number 8,541.23?

   f. tens
   g. thousandths
   h. tenths
   j. hundredths
   k. NH

9. What is the numeral for eighteen thousand, thirty-six?

   a. 18,306
   b. 18,253
   c. 18,036
   d. 8,036
   e. NH

10. What is 605.49 rounded to the nearest whole number?

    f. 606
    g. 605
    h. 605.5
    j. 606.5
    k. NH

Directions

Read each question and choose the correct answer. Mark the space for the answer you have chosen. Mark NH if the answer is not here.

1. 572 + 189 =

   a. 751
   b. 759
   c. 761
   d. 762
   e. NH

6. 5,016 − 2,398 =

   f. 2,616
   g. 2,618
   h. 2,620
   j. 2,622
   k. NH

2. 316 + 489 =

   f. 805
   g. 806
   h. 807
   j. 808
   k. NH

7. 8,927 − 3,859 =

   a. 5,048
   b. 5,058
   c. 5,068
   d. 5,168
   e. NH

3. 500 − 125 =

   a. 425
   b. 375
   c. 350
   d. 275
   e. NH

8. 59,485 + 18,573 =

   f. 77,058
   g. 77,948
   h. 77,958
   j. 78,058
   k. NH

4. 859 + 456 =

   f. 1,318
   g. 1,317
   h. 1,316
   j. 1,315
   k. NH

9. 3,005 − 1,866 =

   a. 1,149
   b. 1,139
   c. 1,136
   d. 1,129
   e. NH

5. 604 − 487 =

   a. 117
   b. 116
   c. 114
   d. 113
   e. NH

10. 6,104 − 2,105 =

   f. 3,991
   g. 3,999
   h. 4,001
   j. 4,009
   k. NH

Name _____   Skill: Multiplication of Whole Numbers

Directions
Read each question and choose the correct answer. Mark the space for the answer you have chosen. Mark NH if the answer is not here.

1. 52 x 16 =

   a. 832
   b. 831
   c. 822
   d. 812
   e. NH

6. 501 x 46 =

   f. 23,086
   g. 23,048
   h. 23,046
   j. 23,038
   k. NH

2. 204 x 12 =

   f. 2,438
   g. 2,446
   h. 2,448
   j. 2,458
   k. NH

7. 77 x 77 =

   a. 5,949
   b. 5,939
   c. 5,929
   d. 5,919
   e. NH

3. 58 x 34 =

   a. 1,962
   b. 1,970
   c. 1,972
   d. 1,982
   e. NH

8. 340 x 25 =

   f. 8,700
   g. 8.600
   h. 8,500
   j. 8,400
   k. NH

4. 51 x 962 =

   f. 49,062
   g. 49,072
   h. 49,162
   j. 49,172
   k. NH

9. 327 x 268 =

   a. 87,616
   b. 87,626
   c. 87,636
   d. 87,646
   e. NH

5. 29 x 312 =

   a. 9,048
   b. 9,056
   c. 5,058
   d. 5,068
   e. NH

10. 2,058 x 692 =

   f. 1,424,126
   g. 1,424,136
   h. 1,424,146
   j. 1,424,156
   k. NH

Name _____   Skill: Division of Whole Numbers

Directions

Read each question and choose the correct answer. Mark the space for the answer you have chosen. Mark NH if the answer is not here.

1. 1,196 ÷ 26 =

   a. 46
   b. 48
   c. 50
   d. 52
   e. NH

6. 648 ÷ 12 =

   f. 44
   g. 52
   h. 53
   j. 54
   k. NH

2. 525 ÷ 15 =

   f. 33
   g. 34
   h. 35
   j. 36
   k. NH

7. 3,000 ÷ 24 =

   a. 110
   b. 115
   c. 120
   d. 125
   e. NH

3. 3,640 ÷ 5 =

   a. 698
   b. 709
   c. 718
   d. 728
   e. NH

8. 4,887 ÷ 9 =

   f. 541
   g. 543
   h. 545
   j. 547
   k. NH

4. 732 ÷ 12 =

   f. 61
   g. 62
   h. 63
   j. 64
   k. NH

9. 3,250 ÷ 13 =

   a. 240
   b. 250
   c. 260
   d. 270
   e. NH

5. 396 ÷ 11 =

   a. 35
   b. 36
   c. 37
   d. 38
   e. NH

10. 4,112 ÷ 16 =

   f. 237
   g. 247
   h. 249
   j. 257
   k. NH

Directions

Read each question and choose the correct answer. Mark the space for the answer you have chosen. Mark NH if the answer is not here.

1. 19 x 20 is between which numbers?

   a. 300 and 400
   b. 400 and 500
   c. 500 and 600
   d. 600 and 700
   e. NH

6. 810 ÷ 40 is between which numbers?

   f. 5 and 10
   g. 10 and 15
   h. 15 and 20
   j. 20 and 30
   k. NH

2. 11 x 50 is between which numbers?

   f. 300 and 400
   g. 400 and 500
   h. 500 and 600
   j. 600 and 700
   k. NH

7. 620 ÷ 20 is between which numbers?

   a. 30 and 40
   b. 40 and 50
   c. 50 and 60
   d. 60 and 70
   e. NH

3. 38 x 30 is between which numbers?

   a. 600 and 700
   b. 1,100 and 1,200
   c. 1,200 and 1,300
   d. 1,500 and 1,600
   e. NH

8. 1,530 ÷ 30 is between which numbers?

   f. 30 and 40
   g. 40 and 50
   h. 50 and 60
   j. 60 and 70
   k. NH

4. 51 x 40 is between which numbers?

   f. 2,000 and 2,100
   g. 1,900 and 1,800
   h. 1,800 and 1,700
   j. 1,700 and 1,600
   k. NH

9. 1,220 ÷ 20 is between which numbers?

   a. 50 and 60
   b. 60 and 70
   c. 70 and 80
   d. 80 and 90
   e. NH

5. 21 x 60 is between which numbers?

   a. 900 and 1,000
   b. 1,000 and 1,100
   c. 1,200 and 1,300
   d. 1,500 and 1,600
   e. NH

10. 564 ÷ 10 is between which numbers?

   f. 50 and 60
   g. 60 and 70
   h. 70 and 80
   j. 80 and 90
   k. NH

Name _____

Directions

Read each question and choose the correct answer. Mark the space for the answer you have chosen. Mark NH if the answer is not here.

---

1. 3.4 + 8.2 =

    a. **11.6**
    b. **11.7**
    c. **11.8**
    d. **11.9**
    e. **NH**

6. 0.4 + 0.5 + 0.9 =

    f. **0.18**
    g. **1.7**
    h. **1.8**
    j. **1.9**
    k. **NH**

---

2. 3.15 + 2.57 =

    f. **5.70**
    g. **5.71**
    h. **5.72**
    j. **5.73**
    k. **NH**

7. $6.24 + $9.58 =

    a. **$15.81**
    b. **$15.82**
    c. **$15.83**
    d. **$15.84**
    e. **NH**

---

3. 57.159 + 14.206 =

    a. **71.364**
    b. **71.365**
    c. **71.366**
    d. **71.367**
    e. **NH**

8. 5.1 + 6.48 =

    f. **11.58**
    g. **11.57**
    h. **11.56**
    j. **6.99**
    k. **NH**

---

4. $5.35 + $8.51 =

    f. **$13.83**
    g. **$13.84**
    h. **$13.85**
    j. **$13.86**
    k. **NH**

9. 9.06 + 3.3 =

    a. **9.39**
    b. **12.34**
    c. **12.36**
    d. **12.39**
    e. **NH**

---

5. 2.58 + 6.18 =

    a. **8.76**
    b. **8.75**
    c. **8.74**
    d. **8.73**
    e. **NH**

10. 6 + 5.8 =

    f. **11.8**
    g. **11.6**
    h. **6.5**
    j. **6.4**
    k. **NH**

Name _____

Directions
Read each question and choose the correct answer. Mark the space for the answer you have chosen. Mark NH if the answer is not here.

1. $5.9 - 2.7 =$	**6.** $6.4 - 5.73 =$
a. 3.2	f. 0.61
b. 3.1	g. 0.63
c. 3.0	h. 0.65
d. 2.9	j. 0.67
e. NH	k. NH
2. $34.89 - 17.94 =$	**7.** $42 - 6.2 =$
f. 16.85	a. 35.7
g. 16.95	b. 35.8
h. 17.05	c. 35.9
j. 17.15	d. 36.20
k. NH	e. NH
3. $60.1 - 37.8 =$	**8.** $37.1 - 29.57 =$
a. 22.2	f. 7.55
b. 22.3	g. 7.54
c. 22.4	h. 7.53
d. 22.5	j. 7.52
e. NH	k. NH
4. $8.461 - 5.289 =$	**9.** $\$6.24 - \$1.50 =$
f. 3.172	a. $4.84
g. 3.171	b. $4.78
h. 3.170	c. $4.76
j. 3.168	d. $4.74
k. NH	e. NH
5. $\$5.49 - \$3.68 =$	**10.** $5 - 4.19 =$
a. $1.85	f. 0.71
b. $1.83	g. 0.81
c. $1.81	h. 0.91
d. $1.79	j. 1.19
e. NH	k. NH

Directions

Read each question and choose the correct answer. Mark the space for the answer you have chosen. Mark NH if the answer is not here.

1. 5.3 x 2 =

   a. 0.106
   b. 1.06
   c. 10.6
   d. 106
   e. NH

6. 0.64 x 1.2 =

   f. 0.0768
   g. 0.768
   h. 7.68
   j. 76.8
   k. NH

2. 81.4 x 6 =

   f. 4.884
   g. 48.84
   h. 488.4
   j. 4,884
   k. NH

7. 5.4 x 6.8 =

   a. 3672
   b. 367.2
   c. 36.72
   d. 3.672
   e. NH

3. 67.25 x 9 =

   a. 605.25
   b. 60.525
   c. 6.0525
   d. 0.60525
   e. NH

8. 13.48 x 1.3 =

   f. 17,524
   g. 1,752.4
   h. 175.24
   j. 17.524
   k. NH

4. 57.33 x 0.2 =

   f. 114.66
   g. 11.466
   h. 1146.6
   j. 11,466
   k. NH

9. 0.09 x 0.001 =

   a. 0.000009
   b. 0.00009
   c. 0.009
   d. 0.09
   e. NH

5. 0.05 x 0.4 =

   a. 2
   b. 0.2
   c. 0.002
   d. 0.02
   e. NH

10. 5.4 x 6.005 =

   f. 32,427
   g. 3,242.7
   h. 324.27
   j. 32.427
   k. NH

Directions
Read each question and choose the correct answer. Mark the space for the answer you have chosen. Mark NH if the answer is not here.

1. $34.8 \div 2 =$

   a. 174
   b. 17.4
   c. 1.74
   d. 0.174
   e. NH

2. $2.79 \div 3 =$

   f. 0.93
   g. 9.3
   h. 93
   j. 930
   k. NH

3. $5.75 \div 5 =$

   a. 0.115
   b. 1.15
   c. 11.5
   d. 115
   e. NH

4. $7.532 \div 4 =$

   f. 1.883
   g. 18.83
   h. 188.3
   j. 1,883
   k. NH

5. $48.6 \div 3 =$

   a. 0.162
   b. 1.62
   c. 16.2
   d. 162
   e. NH

6. $5.6 \div 0.7 =$

   f. 8
   g. 0.8
   h. 0.08
   j. 0.00
   k. NH

7. $1.44 \div 1.2 =$

   a. 12
   b. 1.2
   c. 0.12
   d. 0.012
   e. NH

8. $5.61 \div 0.3 =$

   f. 0.0187
   g. 0.187
   h. 1.87
   j. 18.7
   k. NH

9. $4.095 \div 0.05 =$

   a. 819
   b. 81.9
   c. 8.19
   d. 0.819
   e. NH

10. $4.9 \div 0.007 =$

   f. 700
   g. 70
   h. 7
   j. 0.7
   k. NH

Name _____

Directions
Read each question and choose the correct answer. Mark the space for the answer you have chosen. Mark NH if the answer is not here.

1. 5.3 + 7.26 =	**6.** 6.5 x 98.4 =
a. 12.56 b. 125.6 c. 1,256 d. 12,560 e. NH	f. 638.86 g. 639.6 h. 639.82 j. 640.6 k. NH
2. 2.6 x 5.7 =	**7.** 3.25 ÷ 0.005 =
f. 0.1482 g. 1.482 h. 14.82 j. 1,482 k. NH	a. 650 b. 65 c. 6.5 d. 0.65 e. NH
3. 6.2 – 4.9 =	**8.** 2.5 + 0.31 + 6 =
a. 130 b. 13 c. 1.3 d. 0.13 e. NH	f. 8.81 g. 8.91 h. 8.86 j. 62 k. NH
4. 31.6 ÷ 0.04 =	**9.** 8.05 – 4.99 =
f. 0.79 g. 7.9 h. 79 j. 790 k. NH	a. 0.306 b. 3.06 c. 30.6 d. 306 e. NH
5. 6.4 + 8.28 =	**10.** 64.8 ÷ 2 =
a. 14.68 b. 146.8 c. 1,468 d. 14,680 e. NH	f. 0.0324 g. 0.324 h. 3.24 j. 32.4 k. NH

Directions
Read each question and choose the correct answer. Mark the space for the answer you have chosen. Mark NH if the answer is not here.

---

1. 50.1 ÷ 4.9 is closest to which number?

   a. 7
   b. 8
   c. 9
   d. 10
   e. NH

6. 91 ÷ 29.88 is closest to which number?

   f. 5
   g. 4
   h. 3
   j. 2
   k. NH

---

2. 2.9 x 1.1 is closest to which number?

   f. 2
   g. 3
   h. 4
   j. 5
   k. NH

7. 11.96 x 7.01 is closest to which number?

   a. 77
   b. 80
   c. 84
   d. 87
   e. NH

---

3. 6.89 x 3.09 is closest to which number?

   a. 15
   b. 16
   c. 21
   d. 22
   e. NH

8. 42.008 ÷ 6.93 is closest to which number?

   f. 9
   g. 8
   h. 7
   j. 6
   k. NH

---

4. 35.879 ÷ 5.97 is closest to which number?

   f. 5
   g. 6
   h. 7
   j. 8
   k. NH

9. 11.11 x 9.99 is closest to which number?

   a. 100
   b. 110
   c. 119
   d. 120
   e. NH

---

5. 8.9 x 3.1 is closest to which number?

   a. 24
   b. 27
   c. 30
   d. 32
   e. NH

10. 63.879 ÷ 6.91 is closest to which number?

   f. 9
   g. 10
   h. 11
   j. 12
   k. NH

Name _____   Skill: Addition of Fractions and Mixed Numbers

Directions

Read each question and choose the correct answer. Mark the space for the answer you have chosen. Mark NH if the answer is not here.

1. $\frac{3}{7} + \frac{4}{7} =$

   a. $1\frac{1}{4}$ d. $\frac{1}{2}$

   b. $\frac{12}{7}$ e. NH

   c. $1\frac{5}{7}$

6. $1\frac{5}{6} + 3\frac{2}{9} =$

   f. $5\frac{1}{8}$ j. $5\frac{9}{15}$

   g. $6\frac{1}{9}$ k. NH

   h. $6\frac{1}{18}$

2. $\frac{1}{3} + \frac{3}{4} =$

   f. $\frac{4}{7}$ j. $1\frac{1}{12}$

   g. $\frac{11}{12}$ k. NH

   h. 1

7. $\frac{4}{5} + \frac{8}{9} =$

   a. $\frac{12}{14}$ d. $\frac{12}{13}$

   b. $\frac{6}{7}$ e. NH

   c. $1\frac{31}{45}$

3. $1\frac{1}{4} + 3\frac{1}{6} =$

   a. $4\frac{2}{10}$ d. $4\frac{1}{10}$

   b. $4\frac{5}{12}$ e. NH

   c. $4\frac{1}{5}$

8. $15\frac{7}{12} + 26\frac{3}{8} =$

   f. $41\frac{10}{20}$ j. $42\frac{1}{24}$

   g. $41\frac{1}{2}$ k. NH

   h. $41\frac{23}{24}$

4. $7\frac{3}{8} + 8\frac{3}{8} =$

   f. $15\frac{2}{7}$ j. $15\frac{1}{10}$

   g. $15\frac{2}{10}$ k. NH

   h. $15\frac{3}{4}$

9. $\frac{2}{3} + \frac{1}{4} + \frac{3}{5} =$

   a. $\frac{6}{12}$ d. $\frac{1}{2}$

   b. $1\frac{31}{60}$ e. NH

   c. $1\frac{1}{2}$

5. $\frac{7}{15} + \frac{5}{12} =$

   a. $\frac{12}{27}$ d. $\frac{7}{36}$

   b. $\frac{53}{60}$ e. NH

   c. $\frac{35}{180}$

10. $5\frac{6}{7} + 4\frac{1}{3} =$

   f. $10\frac{4}{21}$ j. $9\frac{7}{10}$

   g. $10\frac{9}{21}$ k. NH

   h. $9\frac{3}{5}$

Directions
Read each question and choose the correct answer. Mark the space for the answer you have chosen. Mark NH if the answer is not here.

---

1. $\frac{2}{3} - \frac{1}{2} =$

   a. 1   d. $\frac{1}{5}$

   b. $\frac{1}{6}$ e. NH

   c. $\frac{1}{2}$

6. $\frac{11}{12} - \frac{3}{8} =$

   f. 2   j. $\frac{13}{24}$

   g. $\frac{11}{24}$ k. NH

   h. $\frac{1}{2}$

---

2. $\frac{7}{9} - \frac{1}{6} =$

   f. $\frac{1}{2}$ j. $\frac{2}{3}$

   g. $\frac{5}{9}$ k. NH

   h. $\frac{11}{18}$

7. $\frac{4}{5} - \frac{2}{7} =$

   a. $\frac{18}{35}$ d. $\frac{3}{7}$

   b. $\frac{17}{35}$ e. NH

   c. $\frac{16}{35}$

---

3. $\frac{3}{4} - \frac{1}{5} =$

   a. $\frac{11}{20}$ d. 2

   b. $\frac{3}{5}$ e. NH

   c. $\frac{13}{20}$

8. $\frac{5}{6} - \frac{3}{8} =$

   f. $\frac{1}{2}$ j. $\frac{11}{24}$

   g. $\frac{13}{14}$ k. NH

   h. $\frac{7}{12}$

---

4. $\frac{3}{5} - \frac{1}{3} =$

   f. $\frac{2}{15}$ j. $\frac{1}{3}$

   g. $\frac{1}{5}$ k. NH

   h. $\frac{4}{15}$

9. $\frac{7}{15} - \frac{1}{12} =$

   a. $\frac{23}{60}$ d. $\frac{13}{30}$

   b. $\frac{2}{5}$ e. NH

   c. $\frac{5}{12}$

---

5. $\frac{4}{5} - \frac{1}{2} =$

   a. $\frac{1}{10}$ d. $\frac{2}{5}$

   b. $\frac{1}{5}$ e. NH

   c. $\frac{3}{10}$

10. $\frac{1}{7} - \frac{1}{8} =$

   f. $\frac{1}{15}$ j. $\frac{1}{55}$

   g. $\frac{1}{56}$ k. NH

   h. $\frac{1}{54}$

Name _____   Skill: Subtraction of Mixed Numbers

Directions

Read each question and choose the correct answer. Mark the space for the answer you have chosen. Mark NH if the answer is not here.

1. $7\frac{3}{4} - 3\frac{1}{2} =$

   a. $4\frac{1}{2}$ d. $3\frac{1}{4}$

   b. $4\frac{1}{4}$ e. NH

   c. $3\frac{1}{2}$

6. $12 - 8\frac{2}{3} =$

   f. $4\frac{1}{3}$ j. $3\frac{2}{3}$

   g. $3\frac{1}{3}$ k. NH

   h. $4\frac{2}{3}$

2. $11\frac{3}{8} - 6\frac{1}{4} =$

   f. $6\frac{1}{8}$ j. $5\frac{1}{2}$

   g. $5\frac{1}{8}$ k. NH

   h. $4\frac{1}{8}$

7. $4\frac{1}{4} - 2\frac{3}{8} =$

   a. $2\frac{1}{8}$ d. $1\frac{1}{2}$

   b. $1\frac{7}{8}$ e. NH

   c. $1\frac{5}{7}$

3. $9\frac{7}{12} - 4\frac{3}{8} =$

   a. $5\frac{4}{24}$ d. $4\frac{5}{24}$

   b. $5\frac{5}{24}$ e. NH

   c. $4\frac{4}{25}$

8. $42\frac{1}{3} - 26\frac{5}{6} =$

   f. $15\frac{1}{2}$ j. $16\frac{1}{3}$

   g. $15\frac{1}{3}$ k. NH

   h. $16\frac{1}{2}$

4. $8\frac{1}{2} - 3\frac{3}{4} =$

   f. $4\frac{3}{4}$ j. $5\frac{3}{4}$

   g. $4\frac{1}{2}$ k. NH

   h. $5\frac{1}{4}$

9. $1\frac{5}{12} - \frac{1}{2} =$

   a. $1\frac{1}{12}$ d. $\frac{5}{6}$

   b. 1 e. NH

   c. $\frac{11}{12}$

5. $18 - 15\frac{1}{2} =$

   a. $3\frac{1}{2}$ d. $2\frac{1}{2}$

   b. $3\frac{1}{4}$ e. NH

   c. $2\frac{3}{4}$

10. $5\frac{1}{2} - 3\frac{5}{6} =$

   f. $1\frac{2}{3}$ j. $2\frac{1}{3}$

   g. $1\frac{1}{3}$ k. NH

   h. $1\frac{1}{2}$

Directions
Read each question and choose the correct answer. Mark the space for the answer you have chosen. Mark NH if the answer is not here.

---

1. $\frac{2}{3}$ x $\frac{9}{10}$ =

   a. $\frac{11}{15}$ d. $\frac{5}{6}$

   b. $\frac{3}{5}$ e. NH

   c. $\frac{4}{5}$

6. $\frac{7}{8}$ x $\frac{2}{3}$ =

   f. $\frac{1}{2}$ j. $\frac{3}{4}$

   g. $\frac{7}{12}$ k. NH

   h. $\frac{2}{3}$

---

2. $\frac{9}{10}$ x $\frac{15}{18}$ =

   f. $\frac{1}{5}$ j. $\frac{3}{4}$

   g. $\frac{1}{4}$ k. NH

   h. $\frac{1}{2}$

7. $\frac{5}{6}$ x $\frac{3}{10}$ =

   a. $\frac{1}{4}$ d. $\frac{1}{5}$

   b. $\frac{1}{3}$ e. NH

   c. $\frac{1}{2}$

---

3. $\frac{6}{7}$ x $\frac{14}{15}$ =

   a. $\frac{4}{5}$ d. $1\frac{1}{5}$

   b. $\frac{3}{5}$ e. NH

   c. $\frac{2}{3}1$

8. $\frac{3}{8}$ x $\frac{6}{7}$ =

   f. $\frac{9}{28}$ j. $\frac{11}{28}$

   g. $\frac{4}{7}$ k. NH

   h. $\frac{3}{7}$

---

4. $\frac{1}{2}$ x $\frac{1}{4}$ =

   f. $\frac{1}{4}$ j. $\frac{2}{6}$

   g. $\frac{1}{8}$ k. NH

   h. $\frac{1}{3}$

9. $\frac{2}{3}$ x $\frac{1}{4}$ =

   a. $\frac{2}{7}$ d. $\frac{1}{6}$

   b. $\frac{3}{7}$ e. NH

   c. $\frac{1}{9}$

---

5. $\frac{3}{5}$ x $\frac{4}{7}$ =

   a. $\frac{7}{12}$ d. $\frac{3}{4}$

   b. $\frac{1}{2}$ e. NH

   c. $\frac{12}{35}$

10. $\frac{5}{6}$ x $\frac{1}{3}$ =

   f. $\frac{6}{7}$ j. $\frac{1}{3}$

   g. $\frac{5}{18}$ k. NH

   h. $\frac{2}{3}$

Directions
Read each question and choose the correct answer. Mark the space for the answer you have chosen. Mark NH if the answer is not here.

1. $3\frac{1}{2} \times 4 =$

   a. $\frac{1}{3}$ d. $\frac{5}{6}$

   b. $\frac{1}{2}$ e. NH

   c. 14

2. $2\frac{1}{3} \times \frac{6}{7} =$

   f. 2   j. $2\frac{6}{21}$

   g. $\frac{1}{2}$ k. NH

   h. $\frac{2}{3}$

3. $2\frac{1}{2} \times \frac{1}{3} =$

   a. $2\frac{1}{10}$ d. $\frac{5}{6}$

   b. $\frac{5}{7}$ e. NH

   c. $\frac{1}{3}$

4. $1\frac{1}{2} \times 2\frac{1}{2} =$

   f. $3\frac{3}{4}$ j. $4\frac{3}{4}$

   g. $2\frac{1}{4}$ k. NH

   h. $4\frac{1}{2}$

5. $2\frac{1}{3} \times \frac{3}{7} =$

   a. $\frac{2}{7}$ d. $2\frac{1}{7}$

   b. 1   e. NH

   c. $1\frac{1}{3}$

6. $3\frac{3}{4} \times 1\frac{3}{5} =$

   f. $3\frac{9}{20}$ j. 8

   g. 6   k. NH

   h. 7

7. $7\frac{1}{2} \times 5\frac{1}{3} =$

   a. $35\frac{6}{7}$ d. 40

   b. $36\frac{5}{18}$ e. NH

   c. $37\frac{2}{3}$

8. $1\frac{5}{6} \times 1\frac{1}{2} =$

   f. $1\frac{5}{12}$ j. $3\frac{1}{4}$

   g. $2\frac{1}{4}$ k. NH

   h. $2\frac{3}{4}$

9. $4\frac{3}{8} \times 3\frac{3}{7} =$

   a. $16\frac{9}{56}$ d. $12\frac{9}{56}$

   b. 15   e. NH

   c. $14\frac{3}{36}$

10. $1\frac{5}{8} \times 1\frac{3}{13} =$

   f. $1\frac{15}{104}$ j. $2\frac{7}{24}$

   g. 2   k. NH

   h. $2\frac{1}{3}$

Directions
Read each question and choose the correct answer. Mark the space for the answer you have chosen. Mark NH if the answer is not here.

---

1. $\frac{3}{7} \div 2 =$

   a. 1   **d.** $\frac{1}{2}$

   b. 2   **e.** NH

   c. $\frac{1}{4}$

6. $\frac{3}{5} \div \frac{4}{5} =$

   f. $\frac{3}{4}$ **j.** $1\frac{1}{3}$

   g. $\frac{12}{25}$ **k.** NH

   h. $1\frac{1}{4}$

---

2. $\frac{3}{4} \div \frac{1}{8} =$

   f. $\frac{3}{4}$ **j.** $\frac{1}{6}$

   g. 6   **k.** NH

   h. 2

7. $\frac{8}{9} \div \frac{1}{3} =$

   a. $2\frac{2}{3}$ **d.** $\frac{8}{12}$

   b. $\frac{3}{8}$ **e.** NH

   c. $2\frac{3}{4}$

---

3. $\frac{2}{3} \div \frac{1}{2} =$

   a. $1\frac{1}{3}$ **d.** $\frac{2}{3}$

   b. $\frac{2}{6}$ **e.** NH

   c. $\frac{3}{4}$

8. $\frac{3}{8} \div \frac{1}{4} =$

   f. $\frac{2}{3}$ **j.** $\frac{3}{32}$

   g. 2   **k.** NH

   h. $1\frac{1}{2}$

---

4. $\frac{1}{5} \div \frac{1}{2} =$

   f. $\frac{1}{5}$ **j.** $2\frac{1}{2}$

   g. $\frac{2}{5}$ **k.** NH

   h. 5

9. $\frac{7}{12} \div \frac{7}{8} =$

   a. $1\frac{1}{2}$ **d.** $1\frac{1}{3}$

   b. $\frac{3}{4}$ **e.** NH

   c. $\frac{2}{3}$

---

5. $\frac{3}{4} \div \frac{1}{3} =$

   a. $\frac{3}{12}$ **d.** $2\frac{1}{4}$

   b. $\frac{1}{4}$ **e.** NH

   c. $2\frac{1}{3}$

10. $\frac{1}{2} \div \frac{3}{8} =$

   f. $1\frac{1}{3}$ **j.** $\frac{3}{16}$

   g. $\frac{3}{4}$ **k.** NH

   h. $\frac{2}{3}$

Name _____   Skill: Division of Mixed Numbers

Directions
Read each question and choose the correct answer. Mark the space for the answer you have chosen. Mark NH if the answer is not here.

1. $1\frac{3}{4} \div \frac{7}{8} =$

    a. $2\frac{3}{4}$ d. $1\frac{17}{32}$

    b. $2\frac{1}{2}$ e. NH

    c. 2

6. $3\frac{3}{5} \div 2\frac{1}{4} =$

    f. $1\frac{2}{5}$ j. 2

    g. $1\frac{3}{5}$ k. NH

    h. $1\frac{4}{5}$

2. $1\frac{2}{3} \div \frac{5}{6} =$

    f. 2   j. $1\frac{1}{4}$

    g. $1\frac{4}{5}$ k. NH

    h. $1\frac{1}{3}$

7. $1\frac{3}{4} \div 1\frac{1}{2} =$

    a. $1\frac{1}{2}$ d. 1

    b. $1\frac{1}{3}$ e. NH

    c. $1\frac{1}{6}$

3. $1\frac{1}{7} \div 1\frac{3}{4} =$

    a. $\frac{28}{49}$ d. 2

    b. $\frac{32}{49}$ e. NH

    c. $1\frac{7}{8}$

8. $6\frac{3}{4} \div 1\frac{7}{8} =$

    f. $3\frac{3}{5}$ j. $2\frac{4}{5}$

    g. $3\frac{2}{5}$ k. NH

    h. $3\frac{1}{5}$

4. $6 \div 1\frac{1}{2} =$

    f. $1\frac{1}{3}$ j. 4

    g. $2\frac{1}{3}$ k. NH

    h. $2\frac{1}{2}$

9. $4\frac{1}{2} \div 1\frac{4}{5} =$

    a. $2\frac{1}{10}$ d. $2\frac{1}{2}$

    b. $2\frac{1}{8}$ e. NH

    c. $2\frac{1}{4}$

5. $7\frac{1}{2} \div 1\frac{2}{3} =$

    a. $3\frac{1}{4}$ d. $4\frac{1}{2}$

    b. $3\frac{1}{2}$ e. NH

    c. $\frac{3}{4}3$

10. $2\frac{1}{2} \div 1\frac{3}{4} =$

    f. $1\frac{3}{7}$ j. $1\frac{1}{2}$

    g. $1\frac{3}{8}$ k. NH

    h. $1\frac{1}{4}$

Directions
Read each question and choose the correct answer. Mark the space for the answer you have
chosen. Mark NH if the answer is not here.

---

1. What decimal equals the fraction $\frac{3}{4}$?

    a. 0.25
    b. 0.5
    c. 0.7
    d. 0.75
    e. NH

6. What decimal equals the fraction $\frac{4}{16}$?

    f. 0.25
    g. 0.40
    h. 0.164
    j. 4.16
    k. NH

---

2. What decimal equals the fraction $\frac{20}{25}$?

    f. 0.75
    g. 0.8
    h. 0.83
    j. 0.9
    k. NH

7. What decimal equals the fraction $\frac{3}{8}$?

    a. 3.8
    b. 0.375
    c. 0.125
    d. 0.525
    e. NH

---

3. What decimal equals the fraction $\frac{1}{6}$?

    a. $0.1\overline{6}$
    b. 0.2
    c. $0.3\overline{6}$
    d. 0.376
    e. NH

8. What decimal equals the fraction $\frac{2}{5}$?

    f. 0.2
    g. 0.3
    h. 0.4
    j. 0.5
    k. NH

---

4. What decimal equals the fraction $\frac{9}{15}$?

    f. 0.8
    g. 0.75
    h. 0.65
    j. 0.6
    k. NH

9. What decimal equals the fraction $\frac{3}{20}$?

    a. 0.15
    b. 0.30
    c. 0.45
    d. 0.60
    e. NH

---

5. What decimal equals the fraction $\frac{5}{100}$?

    a. 0.5
    b. 0.02
    c. 0.20
    d. 0.05
    e. NH

10. What decimal equals the fraction $\frac{9}{50}$?

    f. 0.09
    g. 0.18
    h. 0.27
    j. 0.36
    k. NH

Name _____

Directions

Read each question and choose the correct answer. Mark the space for the answer you have chosen. Mark NH if the answer is not here.

1. What fraction equals the decimal 0.3?

   a. $\frac{1}{3}$ d. $\frac{3}{100}$

   b. $\frac{3}{1}$ e. NH

   c. $\frac{3}{10}$

6. What fraction equals the decimal 8.25?

   f. $8\frac{1}{3}$ j. $8\frac{1}{2}$

   g. $8\frac{1}{4}$ k. NH

   h. $8\frac{2}{5}$

2. What fraction equals the decimal 0.27?

   f. $\frac{2}{7}$ j. $\frac{27}{100}$

   g. $2\frac{7}{10}$ k. NH

   h. $\frac{27}{10}$

7. What fraction equals the decimal 0.35?

   a. $\frac{7}{20}$ d. $\frac{35}{100}$

   b. $\frac{14}{40}$ e. NH

   c. $\frac{17}{50}$

3. What fraction equals the decimal 0.149?

   a. $1\frac{4}{9}$ d. $1\frac{49}{100}$

   b. $1\frac{149}{1,000}$ e. NH

   c. $\frac{149}{100}$

8. What fraction equals the decimal 0.40?

   f. $\frac{4}{100}$ j. $\frac{5}{2}$

   g. $\frac{40}{10}$ k. NH

   h. $\frac{2}{5}$

4. What fraction equals the decimal 1.4?

   f. $1\frac{14}{100}$ j. $1\frac{1}{4}$

   g. $1\frac{4}{100}$ k. NH

   h. $1\frac{2}{5}$

9. What fraction equals the decimal 7.25?

   a. $7\frac{25}{10}$ d. $7\frac{1}{8}$

   b. $7\frac{25}{100}$ e. NH

   c. $7\frac{1}{4}$

5. What fraction equals the decimal 2.5?

   a. $1\frac{5}{10}$ d. $1\frac{1}{20}$

   b. $2\frac{1}{2}$ e. NH

   c. $1\frac{5}{100}$

10. What fraction equals the decimal 0.32?

   f. $\frac{1}{6}$ j. $\frac{4}{50}$

   g. $\frac{8}{23}$ k. NH

   h. $\frac{8}{25}$

Name _____

Directions
Read each question and choose the correct answer. Mark the space for the answer you have chosen. Mark NH if the answer is not here.

1. What is 0.2 written as a percent?

   a. 2%
   b. 22%
   c. 2.2%
   d. 20%
   e. NH

2. What is 0.15 written as a percent?

   f. 1.5%
   g. 15%
   h. 150%
   j. 1500%
   k. NH

3. What is 0.365 written as a percent?

   a. .365%
   b. 3.65%
   c. 36.5%
   d. 365%
   e. NH

4. What is 0.98 written as a percent?

   f. 98%
   g. 9.8%
   h. 0.98%
   j. 0.098%
   k. NH

5. What is 0.01 written as a percent?

   a. 1%
   b. 10%
   c. 100%
   d. 1000%
   e. NH

6. What is 0.39 written as a percent?

   f. 39%
   g. 3.9%
   h. 0.39%
   j. 0.039%
   k. NH

7. What is 0.7 written as a percent?

   a. 7%
   b. 70%
   c. 700%
   d. 0.7%
   e. NH

8. What is 0.06 written as a percent?

   f. 0.6%
   g. 6%
   h. 60%
   j. 600%
   k. NH

9. What is 0.45 written as a percent?

   a. 0.45%
   b. 4.5%
   c. 45%
   d. 450%
   e. NH

10. What is 0.565 written as a percent?

   f. 565%
   g. 56.5%
   h. 5.65%
   j. 0.565%
   k. NH

Directions

Read each question and choose the correct answer. Mark the space for the answer you have chosen. Mark NH if the answer is not here.

1. What is 52% written as a decimal?

  a. 52
  b. 5.2
  c. 0.52
  d. 0.052
  e. NH

6. What is $37\frac{1}{2}$ % written as a decimal?

  f. 37.5
  g. 0.375
  h. 3.75
  j. 375.0
  k. NH

2. What is 6% written as a decimal?

  f. 0.06
  g. 0.6
  h. 6.0
  j. 60.0
  k. NH

7. What is 92% written as a decimal?

  a. 92.0
  b. 9.2
  c. 0.92
  d. 0.092
  e. NH

3. What is 17.5% written as a decimal?

  a. 0.175
  b. 1.75
  c. 17.5
  d. 175.0
  e. NH

8. What is 11% written as a decimal?

  f. 0.11
  g. 1.1
  h. 11.0
  j. 110.0
  k. NH

4. What is 33% written as a decimal?

  f. 33.0
  g. 3.33
  h. 0.333
  j. 0.33
  k. NH

9. What is 36.25% written as a decimal?

  a. 0.03625
  b. 0.3625
  c. 3.625
  d. 36.25
  e. NH

5. What is 45% written as a decimal?

  a. 0.45
  b. 0.045
  c. 4.5
  d. 45.0
  e. NH

10. What is 29% written as a decimal?

  f. 0.029
  g. 0.0029
  h. 0.29
  j. 29.0
  k. NH

Name _____

Directions
Read each question and choose the correct answer. Mark the space for the answer you have chosen. Mark NH if the answer is not here.

---

1. What is $\frac{3}{100}$ written as a percent?

   a. 3%
   b. 30%
   c. 300%
   d. 0.3%
   e. NH

6. What is $\frac{29}{100}$ written as a percent?

   f. 0.29%
   g. 2.9%
   h. 29%
   j. 290%
   k. NH

---

2. What is $\frac{12}{50}$ written as a percent?

   f. 12%
   g. 12.5%
   h. 24%
   j. 2.4%
   k. NH

7. What is $\frac{78}{100}$ written as a percent?

   a. 0.78%
   b. 7.8%
   c. 78%
   d. 780%
   e. NH

---

3. What is $\frac{17}{100}$ written as a percent?

   a. 1.7%
   b. 17%
   c. 170%
   d. 0.17%
   e. NH

8. What is $\frac{1}{2}$ written as a percent?

   f. 50%
   g. 1.2%
   h. 1.5%
   j. 25%
   k. NH

---

4. What is $\frac{91}{100}$ written as a percent?

   f. 0.91%
   g. 9.1%
   h. 910%
   j. 91%
   k. NH

9. What is $\frac{4}{5}$ written as a percent?

   a. 4.5%
   b. 80%
   c. 8%
   d. 800%
   e. NH

---

5. What is $\frac{4}{25}$ written as a percent?

   a. 16%
   b. 4.25%
   c. 8%
   d. 32%
   e. NH

10. What is $\frac{3}{4}$ written as a percent?

   f. 3.4%
   g. 75%
   h. 250%
   j. 80%
   k. NH

Directions
Read each question and choose the correct answer. Mark the space for the answer you have
chosen. Mark NH if the answer is not here.

1. What is 23% written as a fraction in its
   lowest terms?

   a. $\frac{23}{10}$ d. $\frac{23}{100}$

   b. $\frac{230}{100}$ e. NH

   c. $\frac{23}{1}$

6. What is 90% written as a fraction in its
   lowest terms?

   f. $\frac{9}{100}$ j. $\frac{8}{9}$

   g. $\frac{45}{50}$ k. NH

   h. $\frac{9}{10}$

2. What is 49% written as a fraction in its
   lowest terms?

   f. $\frac{49}{100}$ j. $\frac{4}{10}$

   g. $\frac{49}{10}$ k. NH

   h. $\frac{19}{1}$

7. What is 16% written as a fraction in its
   lowest terms?

   a. $\frac{16}{10}$ d. $\frac{4}{25}$

   b. $\frac{8}{10}$ e. NH

   c. $\frac{8}{50}$

3. What is 20% written as a fraction in its
   lowest terms?

   a. $\frac{1}{20}$ d. $\frac{3}{5}$

   b. $\frac{1}{5}$ e. NH

   c. $\frac{2}{5}$

8. What is 35% written as a fraction in its
   lowest terms?

   f. $\frac{7}{20}$ j. $\frac{7}{25}$

   g. $\frac{35}{50}$ k. NH

   h. $\frac{14}{50}$

4. What is 60% written as a fraction in its
   lowest terms?

   f. $\frac{1}{5}$ j. $\frac{4}{5}$

   g. $\frac{2}{5}$ k. NH

   h. $\frac{3}{5}$

9. What is 50% written as a fraction in its
   lowest terms?

   a. $\frac{1}{3}$ d. $\frac{1}{5}$

   b. $\frac{1}{2}$ e. NH

   c. $\frac{1}{4}$

5. What is 75% written as a fraction in its
   lowest terms?

   a. $\frac{3}{4}$ d. $\frac{2}{3}$

   b. $\frac{1}{4}$ e. NH

   c. $\frac{1}{2}$

10. What is 55% written as a fraction in its
    lowest terms?

    f. $\frac{1}{2}$ j. $\frac{8}{9}$

    g. $\frac{11}{20}$ k. NH

    h. $\frac{6}{7}$

Directions

Read each question and choose the correct answer. Mark the space for the answer you have chosen. Mark NH if the answer is not here.

1. What is 0.65 written as a percent?

   a. 6.5%
   b. 65%
   c. 650%
   d. 0.65%
   e. NH

6. What is 0.24 written as a fraction in its lowest terms?

   f. $\frac{8}{50}$ j. $\frac{12}{50}$

   g. $\frac{16}{50}$ k. NH

   h. $\frac{6}{25}$

2. What is 0.42 written as a fraction in its lowest terms?

   f. $\frac{42}{100}$ j. $\frac{21}{50}$

   g. $\frac{15}{60}$ k. NH

   h. $\frac{10}{50}$

7. What is 0.175 written as a percent?

   a. 175%
   b. 1.75%
   c. 1750%
   d. 17.5%
   e. NH

3. What is $\frac{4}{5}$ written as a percent?

   a. 4.5%
   b. 80%
   c. 60%
   d. 45%
   e. NH

8. What is $\frac{2}{5}$ written as a decimal?

   f. 4.0
   g. 0.4
   h. 0.04
   j. 0.004
   k. NH

4. What is 0.3 written as a percent?

   f. 3%
   g. 30%
   h. 300%
   j. 0.3%
   k. NH

9. What is $\frac{1}{4}$ written as a percent?

   a. 14%
   b. 25%
   c. 50%
   d. 1.4%
   e. NH

5. What is $\frac{3}{8}$ written as a decimal?

   a. 3.85
   b. 0.425
   c. 0.375
   d. 0.525
   e. NH

10. What is $\frac{7}{10}$ written as a percent?

   f. 7%
   g. 70%
   h. 700%
   j. 0.7%
   k. NH

Name _____

Directions

Read each question and choose the correct answer. Mark the space for the answer you have chosen. Mark NH if the answer is not here.

1. Estimate the answer by rounding:
$$37 + 56 =$$

a. 100
b. 90
c. 80
d. 70
e. NH

2. Estimate the answer by rounding:
$$52 - 16 =$$

f. 20
g. 30
h. 40
j. 50
k. NH

3. Estimate the answer by rounding:
$$198 + 315 =$$

a. 500
b. 600
c. 700
d. 800
e. NH

4. Estimate the answer by rounding:
$$507 - 182 =$$

f. 700
g. 600
h. 500
j. 400
k. NH

5. Estimate the answer by rounding:
$$1,987 - 1,413 =$$

a. 700
b. 600
c. 500
d. 400
e. NH

6. Estimate the answer by rounding:
$$617 + 486 =$$

f. 1,000
g. 1,100
h. 1,200
j. 1,300
k. NH

7. Estimate the answer by rounding:
$$510 - 398 =$$

a. 400
b. 300
c. 200
d. 100
e. NH

8. Estimate the answer by rounding:
$$206 + 815 + 763 =$$

f. 1,600
g. 1,800
h. 2,000
j. 2,200
k. NH

9. Estimate the answer by rounding:
$$890 - 625 =$$

a. 100
b. 200
c. 300
d. 400
e. NH

10. Estimate the answer by rounding:
$$69 + 52 + 86 + 49 =$$

f. 260
g. 250
h. 240
j. 230
k. NH

Directions

Read each question and choose the correct answer. Mark the space for the answer you have chosen. Mark NH if the answer is not here.

---

1. Estimate the answer by rounding:

   32 x 49 =

   a. 1,400
   b. 1,500
   c. 1,600
   d. 1,700
   e. NH

6. Estimate the answer by rounding:

   719 ÷ 8.9 =

   f. 80
   g. 78
   h. 76
   j. 74
   k. NH

---

2. Estimate the answer by rounding:

   61 x 7.9 =

   f. 420
   g. 440
   h. 460
   j. 480
   k. NH

7. Estimate the answer by rounding:

   52 x 61 =

   a. 3,000
   b. 3,200
   c. 3,600
   d. 3,800
   e. NH

---

3. Estimate the answer by rounding:

   631 ÷ 71 =

   a. 12
   b. 11
   c. 10
   d. 9
   e. NH

8. Estimate the answer by rounding:

   990 ÷ 49 =

   f. 15
   g. 20
   h. 25
   j. 30
   k. NH

---

4. Estimate the answer by rounding:

   510 x 39 =

   f. 16,000
   g. 18,000
   h. 20,000
   j. 22,000
   k. NH

9. Estimate the answer by rounding:

   63.9 ÷ 7.8 =

   a. 5
   b. 6
   c. 7
   d. 8
   e. NH

---

5. Estimate the answer by rounding:

   419 ÷ 6 =

   a. 80
   b. 70
   c. 60
   d. 50
   e. NH

10. Estimate the answer by rounding:

   419 x 59 =

   f. 24,000
   g. 22,000
   h. 20,000
   j. 18,000
   k. NH

Directions
Read each question and choose the correct answer. Mark the space for the answer you have chosen. Mark NH if the answer is not here.

1. What is 10% of 50?

   a. 4
   b. 5
   c. 6
   d. 7
   e. NH

6. What is 60% of 100?

   f. 40
   g. 50
   h. 60
   j. 70
   k. NH

2. What is 25% of 16?

   f. 3
   g. 4
   h. 5
   j. 6
   k. NH

7. What is 20% of 40?

   a. 5
   b. 6
   c. 7
   d. 8
   e. NH

3. What is 50% of 18?

   a. 6
   b. 7
   c. 8
   d. 9
   e. NH

8. What is 25% of 20?

   f. 5
   g. 4
   h. 3
   j. 2
   k. NH

4. What is 75% of 200?

   f. 125
   g. 140
   h. 150
   j. 175
   k. NH

9. What is 10% of 460?

   a. 45
   b. 46
   c. 47
   d. 48
   e. NH

5. What is 10% of 360?

   a. 33
   b. 34
   c. 35
   d. 36
   e. NH

10. What is 50% of 80?

   f. 40
   g. 30
   h. 20
   j. 10
   k. NH

Name _____

Directions
Read each question and choose the correct answer. Mark the space for the answer you have chosen. Mark NH if the answer is not here.

1. 8 is 10% of what number?

   a. 80
   b. 70
   c. 60
   d. 50
   e. NH

6. 9 is 75% of what number?

   f. 6
   g. 8
   h. 15
   j. 12
   k. NH

2. 12 is 25% of what number?

   f. 12
   g. 24
   h. 48
   j. 96
   k. NH

7. 2 is 25% of what number?

   a. 6
   b. 8
   c. 10
   d. 12
   e. NH

3. 6 is 20% of what number?

   a. 30
   b. 25
   c. 24
   d. 18
   e. NH

8. 14 is 50% of what number?

   f. 16
   g. 28
   h. 42
   j. 56
   k. NH

4. 5 is 50% of what number?

   f. 7.5
   g. 10
   h. 20
   j. 25
   k. NH

9. 36 is 10% of what number?

   a. 360
   b. 180
   c. 480
   d. 720
   e. NH

5. 19 is 10% of what number?

   a. 190
   b. 180
   c. 160
   d. 95
   e. NH

10. 40 is 25% of what number?

   f. 120
   g. 140
   h. 160
   j. 200
   k. NH

Name _____

Directions

Read each question and choose the correct answer. Mark the space for the answer you have chosen. Mark NH if the answer is not here.

1. 5 is what percent of 25?

   a. 10%
   b. 20%
   c. 25%
   d. 30%
   e. NH

2. 12 is what percent of 24?

   f. 30%
   g. 40%
   h. 50%
   j. 60%
   k. NH

3. 9 is what percent of 90?

   a. 25%
   b. 20%
   c. 15%
   d. 10%
   e. NH

4. 27 is what percent of 90?

   f. 15%
   g. 20%
   h. 25%
   j. 30%
   k. NH

5. 4 is what percent of 20?

   a. 20%
   b. 25%
   c. 30%
   d. 40%
   e. NH

6. 18 is what percent of 180?

   f. 25%
   g. 20%
   h. 15%
   j. 10%
   k. NH

7. 42 is what percent of 84?

   a. 20%
   b. 40%
   c. 50%
   d. 60%
   e. NH

8. 25 is what percent of 100?

   f. 20%
   g. 25%
   h. 30%
   j. 40%
   k. NH

9. 3 is what percent of 12?

   a. 15%
   b. 20%
   c. 25%
   d. 30%
   e. NH

10. 30 is what percent of 40?

   f. 25%
   g. 50%
   h. 75%
   j. 80%
   k. NH

Directions
Read each question and choose the correct answer. Mark the space for the answer you have chosen. Mark NH if the answer is not here.

1. What number is 10% of 60?

   a. 3
   b. 4
   c. 5
   d. 6
   e. NH

2. 4 is 20% of what number?

   f. 16
   g. 20
   h. 24
   j. 28
   k. NH

3. 12 is what percent of 60?

   a. 20%
   b. 25%
   c. 30%
   d. 40%
   e. NH

4. What number is 50% of 38?

   f. 17
   g. 18
   h. 19
   j. 20
   k. NH

5. 16 is 50% of what number?

   a. 20
   b. 24
   c. 28
   d. 32
   e. NH

6. 15 is what percent of 30?

   f. 20%
   g. 30%
   h. 40%
   j. 50%
   k. NH

7. What number is 30% of 70?

   a. 21
   b. 20
   c. 18
   d. 15
   e. NH

8. What number is 20% of 140?

   f. 26
   g. 28
   h. 30
   j. 32
   k. NH

9. 15 is 25% of what number?

   a. 50
   b. 60
   c. 70
   d. 80
   e. NH

10. 5 is what percent of 50?

   f. 10%
   g. 15%
   h. 20%
   j. 25%
   k. NH

Name _____

Directions
Read each question and choose the correct answer. Mark the space for the answer you have chosen. Mark NH if the answer is not here.

1. If n + 3 = 8, then n =

   a. 2
   b. 3
   c. 4
   d. 5
   e. NH

6. If n + 9 = 17, then n =

   f. 8
   g. 7
   h. 6
   j. 5
   k. NH

2. If n + 5 = 9, then n =

   f. 4
   g. 5
   h. 6
   j. 7
   k. NH

7. If n + 4 = 12, then n =

   a. 8
   b. 7
   c. 6
   d. 5
   e. NH

3. If n + 3 = 12, then n =

   a. 7
   b. 8
   c. 9
   d. 10
   e. NH

8. If n + 6 = 12, then n =

   f. 8
   g. 7
   h. 6
   j. 5
   k. NH

4. If n + 8 = 15, then n =

   f. 6
   g. 7
   h. 8
   j. 9
   k. NH

9. If n + 9 = 15, then n =

   a. 8
   b. 7
   c. 6
   d. 5
   e. NH

5. If n + 6 = 11, then n =

   a. 4
   b. 5
   c. 6
   d. 7
   e. NH

10. If n + 2 = 11, then n =

   f. 9
   g. 8
   h. 7
   j. 6
   k. NH

Directions

Read each question and choose the correct answer. Mark the space for the answer you have chosen. Mark NH if the answer is not here.

1. If $n - 5 = 7$, then $n =$

   a. 11
   b. 12
   c. 13
   d. 14
   e. NH

6. If $n - 2 = 8$, then $n =$

   f. 8
   g. 9
   h. 10
   j. 11
   k. NH

2. If $n - 8 = 6$, then $n =$

   f. 12
   g. 13
   h. 14
   j. 15
   k. NH

7. If $n - 5 = 9$, then $n =$

   a. 12
   b. 13
   c. 14
   d. 15
   e. NH

3. If $n - 4 = 7$, then $n =$

   a. 11
   b. 12
   c. 13
   d. 14
   e. NH

8. If $n - 7 = 7$, then $n =$

   f. 14
   g. 15
   h. 16
   j. 17
   k. NH

4. If $n - 9 = 6$, then $n =$

   f. 18
   g. 17
   h. 16
   j. 15
   k. NH

9. If $n - 9 = 8$, then $n =$

   a. 18
   b. 17
   c. 16
   d. 15
   e. NH

5. If $n - 7 = 2$, then $n =$

   a. 8
   b. 9
   c. 10
   d. 11
   e. NH

10. If $n - 4 = 3$, then $n =$

   f. 7
   g. 6
   h. 5
   j. 4
   k. NH

Name _____

Directions

Read each question and choose the correct answer. Mark the space for the answer you have chosen. Mark NH if the answer is not here.

1. If 3 x n = 18, then n =

   a. 5
   b. 6
   c. 7
   d. 8
   e. NH

2. If 4n = 28, then n =

   f. 5
   g. 6
   h. 7
   j. 8
   k. NH

3. If 8n = 64, then n =

   a. 7
   b. 8
   c. 9
   d. 10
   e. NH

4. If 6n = 54, then n =

   f. 7
   g. 8
   h. 9
   j. 10
   k. NH

5. If 5n = 45, then n =

   a. 7
   b. 8
   c. 9
   d. 10
   e. NH

6. If 12n = 96, then n =

   f. 11
   g. 10
   h. 9
   j. 8
   k. NH

7. If 7n = 63, then n =

   a. 9
   b. 8
   c. 7
   d. 6
   e. NH

8. If 9n = 36, then n =

   f. 7
   g. 6
   h. 5
   j. 4
   k. NH

9. If 3n = 21, then n =

   a. 7
   b. 8
   c. 9
   d. 10
   e. NH

10. If 9n = 108, then n =

   f. 10
   g. 11
   h. 12
   j. 13
   k. NH

Directions
Read each question and choose the correct answer. Mark the space for the answer you have chosen. Mark NH if the answer is not here.

1. If $\frac{n}{3} = 8$, then n =

   a. 18
   b. 21
   c. 24
   d. 27
   e. NH

6. If $\frac{n}{6} = 9$, then n =

   f. 54
   g. 56
   h. 60
   j. 66
   k. NH

2. If $\frac{n}{2} = 4$, then n =

   f. 8
   g. 10
   h. 12
   j. 14
   k. NH

7. If $\frac{n}{5} = 7$, then n =

   a. 25
   b. 30
   c. 40
   d. 45
   e. NH

3. If $\frac{n}{6} = 7$, then n =

   a. 24
   b. 30
   c. 36
   d. 42
   e. NH

8. If $\frac{n}{12} = 3$, then n =

   f. 12
   g. 24
   h. 36
   j. 48
   k. NH

4. If $\frac{n}{4} = 8$, then n =

   f. 32
   g. 28
   h. 24
   j. 20
   k. NH

9. If $\frac{n}{8} = 11$, then n =

   a. 80
   b. 88
   c. 96
   d. 104
   e. NH

5. If $\frac{n}{9} = 7$, then n =

   a. 49
   b. 56
   c. 63
   d. 70
   e. NH

10. If $\frac{n}{3} = 9$, then n =

   f. 27
   g. 24
   h. 18
   j. 12
   k. NH

Directions
Read each question and choose the correct answer. Mark the space for the answer you have chosen. Mark NH if the answer is not here.

1. If n = 4, then n + 2 =

    a. 4
    b. 5
    c. 6
    d. 7
    e. NH

6. If n = 7, then 2n − 9 =

    f. 7
    g. 6
    h. 5
    j. 4
    k. NH

2. If n = 3, then 2n =

    f. 2
    g. 4
    h. 6
    j. 9
    k. NH

7. If n = 6, then 5n + 1 =

    a. 28
    b. 29
    c. 30
    d. 31
    e. NH

3. If n = 8, then 3n + 1 =

    a. 25
    b. 21
    c. 14
    d. 6
    e. NH

8. If n = 12, then 5n + 2 =

    f. 52
    g. 62
    h. 64
    j. 72
    k. NH

4. If n = 2, then 8n + 4 =

    f. 14
    g. 16
    h. 18
    j. 20
    k. NH

9. If n = 10, then 8n =

    a. 18
    b. 60
    c. 80
    d. 90
    e. NH

5. If n = 4, then 2n + 7 =

    a. 14
    b. 15
    c. 16
    d. 17
    e. NH

10. If n = 9, then n + 7 =

    f. 16
    g. 15
    h. 14
    j. 13
    k. NH

Directions

Read each question and choose the correct answer. Mark the space for the answer you have chosen. Mark NH if the answer is not here.

1. If n = 4, then 4n + 6 =

    a. 14
    b. 20
    c. 22
    d. 24
    e. NH

6. If n = 12, then 7n + 8 =

    f. 92
    g. 88
    h. 84
    j. 82
    k. NH

2. If n = 8, then 2n − 3 =

    f. 11
    g. 12
    h. 13
    j. 14
    k. NH

7. If n = 9, then 5n + 5 =

    a. 40
    b. 45
    c. 50
    d. 55
    e. NH

3. If n = 15, then n − 8 =

    a. 6
    b. 7
    c. 8
    d. 9
    e. NH

8. If n = 100, then 7n + 40 =

    f. 704
    g. 740
    h. 744
    j. 784
    k. NH

4. If n = 20, then 3n + 2 =

    f. 25
    g. 60
    h. 62
    j. 64
    k. NH

9. If n = 23, then 2n − 6 =

    a. 40
    b. 36
    c. 34
    d. 32
    e. NH

5. If n = 18, then 5n − 3 =

    a. 97
    b. 90
    c. 87
    d. 80
    e. NH

10. If n = 35, then 3n + 4 =

    f. 109
    g. 106
    h. 103
    j. 100
    k. NH

Directions
Read each question and choose the correct answer. Mark the space for the answer you have
chosen. Mark NH if the answer is not here.

1. If $\frac{x}{2} = \frac{4}{8}$, then x =

   a. 0
   b. 1
   c. 2
   d. 3
   e. NH

6. If $\frac{5}{x} = \frac{30}{36}$, then x =

   f. 3
   g. 4
   h. 5
   j. 6
   k. NH

2. If $\frac{3}{8} = \frac{6}{x}$, then x =

   f. 10
   g. 12
   h. 13
   j. 16
   k. NH

7. If $\frac{3}{9} = \frac{x}{27}$, then x =

   a. 8
   b. 9
   c. 10
   d. 11
   e. NH

3. If $\frac{3}{5} = \frac{x}{15}$, then x =

   a. 6
   b. 9
   c. 12
   d. 13
   e. NH

8. If $\frac{2}{6} = \frac{6}{x}$, then x =

   f. 16
   g. 18
   h. 24
   j. 30
   k. NH

4. If $\frac{9}{x} = \frac{18}{20}$, then x =

   f. 14
   g. 16
   h. 18
   j. 20
   k. NH

9. If $\frac{4}{5} = \frac{x}{35}$, then x =

   a. 7
   b. 28
   c. 35
   d. 42
   e. NH

5. If $\frac{3}{4} = \frac{x}{16}$, then x =

   a. 12
   b. 14
   c. 16
   d. 18
   e. NH

10. If $\frac{5}{7} = \frac{10}{x}$, then x =

   f. 13
   g. 14
   h. 15
   j. 16
   k. NH

Directions

Read each question and choose the correct answer. Mark the space for the answer you have chosen. Mark NH if the answer is not here.

---

1. If $\frac{2}{6} = \frac{x}{9}$, then x =

   a. 3
   b. 4
   c. 5
   d. 6
   e. NH

6. If $\frac{4}{20} = \frac{3}{x}$, then x =

   f. 13
   g. 14
   h. 15
   j. 16
   k. NH

---

2. If $\frac{6}{x} = \frac{15}{20}$, then x =

   f. 5
   g. 6
   h. 7
   j. 8
   k. NH

7. If $\frac{15}{x} = \frac{10}{12}$, then x =

   a. 16
   b. 18
   c. 21
   d. 24
   e. NH

---

3. If $\frac{x}{9} = \frac{5}{15}$, then x =

   a. 2
   b. 3
   c. 4
   d. 5
   e. NH

8. If $\frac{6}{8} = \frac{x}{28}$, then x =

   f. 18
   g. 21
   h. 24
   j. 27
   k. NH

---

4. If $\frac{5}{10} = \frac{x}{6}$, then x =

   f. 1
   g. 2
   h. 3
   j. 4
   k. NH

9. If $\frac{20}{x} = \frac{8}{12}$, then x =

   a. 24
   b. 27
   c. 28
   d. 30
   e. NH

---

5. If $\frac{3}{12} = \frac{24}{x}$, then x =

   a. 84
   b. 96
   c. 108
   d. 112
   e. NH

10. If $\frac{x}{15} = \frac{2}{6}$, then x =

   f. 5
   g. 6
   h. 7
   j. 8
   k. NH

Directions
Read each question and choose the correct answer. Mark the space for the answer you have chosen. Mark NH if the answer is not here.

---

1. Estimate the answer by rounding:
$$48.9 \div 6.95 =$$

  a. 6
  b. 7
  c. 8
  d. 9
  e. NH

6. 15 x 22 is between which numbers?

  f. **100 and 200**
  g. **200 and 300**
  h. **300 and 400**
  j. **400 and 500**
  k. **NH**

---

2. What is 0.26 written as a fraction in its lowest terms?

  f. $\frac{26}{10}$ j. $\frac{13}{50}$

  g. $\frac{26}{100}$ k. **NH**

  h. $\frac{14}{50}$

7. $12.34 - 8.96 =$

  a. **3.13**
  b. **3.28**
  c. **3.38**
  d. **3.48**
  e. **NH**

---

3. $1,586 + 9,144 =$

  a. **10,720**
  b. **10,730**
  c. **10,740**
  d. **10,750**
  e. **NH**

8. 3 is what percent of 6?

  f. **50%**
  g. **25%**
  h. **15%**
  j. **10%**
  k. **NH**

---

4. Estimate the answer by rounding:
$$98 + 52 =$$

  f. **130**
  g. **140**
  h. **150**
  j. **160**
  k. **NH**

9. What is $\frac{3}{5}$ written as a percent?

  a. **35%**
  b. **3.5%**
  c. **60%**
  d. **80%**
  e. **NH**

---

5. $5\frac{1}{3} \times \frac{3}{4} =$

  a. **4**   d. $5\frac{1}{4}$

  b. $4\frac{1}{4}$ e. **NH**

  c. $4\frac{1}{3}$

10. $2,832 \div 48 =$

  f. **58**
  g. **59**
  h. **60**
  j. **61**
  k. **NH**

Directions

Read each question and choose the correct answer. Mark the space for the answer you have chosen. Mark NH if the answer is not here.

1. Estimate the answer by rounding:
$$35.8 \div 8.99 =$$

   a. 5
   b. 4
   c. 6
   d. 7
   e. NH

2. What is 0.31 written as a fraction in its lowest terms?

   f. $\frac{31}{10}$ j. $\frac{31}{30}$

   g. $\frac{31}{100}$ k. NH

   h. $\frac{12}{45}$

3. $2,341 + 8,711 =$

   a. 11,622
   b. 11,205
   c. 11,052
   d. 11,502
   e. NH

4. Estimate the answer by rounding:
$$61 + 32 =$$

   f. 88
   g. 93
   h. 74
   j. 86
   k. NH

5. $4\frac{1}{5} \times \frac{5}{7} =$

   a. 3   d. $3\frac{1}{4}$

   b. $3\frac{1}{4}$ e. NH

   c. $3\frac{1}{3}$

6. 10 x 48 is between which numbers?

   f. 100 and 200
   g. 200 and 300
   h. 300 and 400
   j. 400 and 500
   k. NH

7. $11.21 - 6.58 =$

   a. 4.13
   b. 4.63
   c. 4.36
   d. 4.33
   e. NH

8. 5 is what percent of 20?

   f. 10%
   g. 25%
   h. 35%
   j. 20%
   k. NH

9. What is $\frac{2}{4}$ written as a percent?

   a. 24%
   b. 2.4%
   c. 50%
   d. 40%
   e. NH

10. $976 \div 16 =$

   f. 38
   g. 54
   h. 60
   j. 61
   k. NH

Directions

Read each question and choose the correct answer. Mark the space for the answer you have chosen. Mark NH if the answer is not here.

1. Estimate the answer by rounding:

   $$53.999 \div 6.25 =$$

   a. 9
   b. 6
   c. 8
   d. 7
   e. NH

2. What is 0.13 written as a fraction in its lowest terms?

   f. $\frac{13}{1000}$ j. $\frac{13}{12}$

   g. $\frac{13}{100}$ k. NH

   h. $\frac{13}{10}$

3. $2,414 + 8,213 =$

   a. 10,670
   b. 10,627
   c. 10,672
   d. 10,762
   e. NH

4. Estimate the answer by rounding:

   $$44 + 31 =$$

   f. 70
   g. 60
   h. 80
   j. 50
   k. NH

5. $2\frac{1}{6} \times \frac{6}{8} =$

   a. 2   d. $1\frac{5}{8}$

   b. $1\frac{1}{2}$ e. NH

   c. $1\frac{1}{6}$

6. 12 x 41 is between which numbers?

   f. 100 and 200
   g. 200 and 300
   h. 300 and 400
   j. 400 and 500
   k. NH

7. $18.21 - 4.32 =$

   a. 13.23
   b. 13.80
   c. 13.89
   d. 13.98
   e. NH

8. 9 is what percent of 27?

   f. 40%
   g. 20%
   h. 30%
   j. 15%
   k. NH

9. What is $\frac{1}{4}$ written as a percent?

   a. 1.14%
   b. 1.4%
   c. 25%
   d. 20%
   e. NH

10. $1,872 \div 39 =$

    f. 38
    g. 52
    h. 48
    j. 49
    k. NH

Name _____   Skill: Computation Practice

Directions
Read each question and choose the correct answer. Mark the space for the answer you have chosen. Mark NH if the answer is not here.

---

1. Estimate the answer by rounding:
$$19.78 \div 4.75 =$$
   a. 7
   b. 4
   c. 5
   d. 8
   e. NH

6. 15 x 21 is between which numbers?
   f. 100 and 200
   g. 200 and 300
   h. 300 and 400
   j. 400 and 500
   k. NH

---

2. What is 0.87 written as a fraction in its lowest terms?
   f. $\frac{87}{10}$ j. $\frac{87}{40}$
   g. $\frac{87}{100}$ k. NH
   h. $\frac{45}{50}$

7. $15.48 - 9.85 =$
   a. 5.36
   b. 5.63
   c. 5.31
   d. 5.35
   e. NH

---

3. $3,982 + 8,234 =$
   a. 12,126
   b. 12,211
   c. 12,216
   d. 12,612
   e. NH

8. 8 is what percent of 24?
   f. 40%
   g. 30%
   h. 25%
   j. 15%
   k. NH

---

4. Estimate the answer by rounding:
$$43 + 62 =$$
   f. 120
   g. 100
   h. 110
   j. 130
   k. NH

9. What is $\frac{4}{5}$ written as a percent?
   a. 45%
   b. 4.5%
   c. 75%
   d. 80%
   e. NH

---

5. $4\frac{1}{7} \times \frac{7}{8} =$
   a. 3   d. $4\frac{1}{7}$
   b. $3\frac{1}{7}$ e. NH
   c. $3\frac{5}{8}$

10. $2,698 \div 71 =$
   f. 41
   g. 38
   h. 39
   j. 29
   k. NH

Name _____   Skill: Computation Practice

Directions
Read each question and choose the correct answer. Mark the space for the answer you have chosen. Mark NH if the answer is not here.

---

1. Estimate the answer by rounding:
    $$37.18 \div 8.71 =$$

    a. 5
    b. 4
    c. 7
    d. 6
    e. NH

6. 9 x 38 is between which numbers?

    f. 100 and 200
    g. 200 and 300
    h. 300 and 400
    j. 400 and 500
    k. NH

---

2. What is 0.17 written as a fraction in its lowest terms?

    f. $\frac{17}{10}$ j. $\frac{17}{35}$

    g. $\frac{17}{100}$ k. NH

    h. $\frac{17}{50}$

7. $14.55 - 7.87 =$

    a. 6.13
    b. 6.65
    c. 6.67
    d. 6.68
    e. NH

---

3. $2,722 + 6,521 =$

    a. 9,993
    b. 9,122
    c. 9,321
    d. 9,243
    e. NH

8. 4 is what percent of 12?

    f. 40%
    g. 15%
    h. 30%
    j. 20%
    k. NH

---

4. Estimate the answer by rounding:
    $$60 + 22 =$$

    f. 80
    g. 90
    h. 70
    j. 85
    k. NH

9. What is $\frac{2}{4}$ written as a percent?

    a. 50%
    b. 2.4%
    c. 24%
    d. 42%
    e. NH

---

5. $5\frac{1}{3} \times \frac{3}{4} =$

    a. 2   d. $1\frac{5}{9}$

    b. $1\frac{7}{9}$ e. NH

    c. $1\frac{8}{9}$

10. $1,836 \div 36 =$

    f. 37
    g. 41
    h. 50
    j. 51
    k. NH

Directions

Read each question and choose the correct answer. Mark the space for the answer you have chosen. Mark NH if the answer is not here.

---

1. $7\frac{1}{3} + 8\frac{1}{4} =$

  a. $15\frac{2}{7}$ **d.** $15\frac{3}{4}$

  b. $15\frac{1}{7}$ **e.** NH

  c. $15\frac{7}{12}$

6. What is 0.13 written as a percent?

  f. 1.3%
  g. 130%
  h. 13%
  j. 0.13%
  k. NH

---

2. If $n + 3 = 12$, then $n =$

  f. 9
  g. 10
  h. 11
  j. 12
  k. NH

7. If $\frac{3}{4} = \frac{15}{n}$, then $n =$

  a. 15
  b. 20
  c. 25
  d. 30
  e. NH

---

3. What is $\frac{3}{8}$ written as a decimal?

  a. 3.8
  b. 0.308
  c. 0.375
  d. 2.75
  e. NH

8. $37 \times 52 =$

  f. 1,924
  g. 1,926
  h. 1,934
  j. 1,942
  k. NH

---

4. If $n = 3$, then $8n - 4 =$

  f. 23
  g. 22
  h. 21
  j. 20
  k. NH

9. $\frac{8}{9} \div \frac{1}{3} =$

  a. $\frac{2}{3}$ **d.** $\frac{3}{8}$

  b. $2\frac{2}{3}$ **e.** NH

  c. $2\frac{5}{8}$

---

5. Estimate the answer by rounding:
$$9.6 \times 3.8 =$$

  a. 40
  b. 30
  c. 20
  d. 10
  e. NH

10. What is 52% written as a fraction in its lowest terms?

  f. $\frac{13}{25}$ **j.** $\frac{52}{100}$

  g. $\frac{26}{50}$ **k.** NH

  h. $\frac{52}{10}$

Directions
Read each question and choose the correct answer. Mark the space for the answer you have chosen. Mark NH if the answer is not here.

---

1. $1\frac{1}{2} \div 1\frac{2}{3} =$

   a. $\frac{4}{5}$ d. $2\frac{1}{2}$

   b. $\frac{9}{10}$ e. NH

   c. $\frac{3}{5}$

6. Estimate the answer by rounding:
$$47.9 \div 5.9 =$$

   f. 6
   g. 7
   h. 8
   j. 9
   k. NH

---

2. Estimate the answer by rounding:
$$57 \times 39 =$$

   f. 2,300
   g. 2,400
   h. 2,500
   j. 2,600
   k. NH

7. What is $\frac{3}{4}$ written as a decimal?

   a. 0.34
   b. 0.75
   c. 7.5
   d. 3.4
   e. NH

---

3. What is 0.63 written as a percent?

   a. 63%
   b. 603%
   c. 6.3%
   d. 0.63%
   e. NH

8. $6,001 - 5,289 =$

   f. 728
   g. 722
   h. 718
   j. 712
   k. NH

---

4. $509 \times 15 =$

   f. 7,635
   g. 7,645
   h. 7,655
   j. 7,665
   k. NH

9. Estimate the answer by rounding:
$$3.99 \times 4.01 =$$

   a. 13
   b. 14
   c. 15
   d. 16
   e. NH

---

5. If $\frac{3}{n} = \frac{9}{15}$, then n =

   a. 4
   b. 5
   c. 6
   d. 7
   e. NH

10. What is 60% of 50?

   f. 25
   g. 30
   h. 35
   j. 40
   k. NH

Directions

Read each question and choose the correct answer. Mark the space for the answer you have chosen. Mark NH if the answer is not here.

---

1. What is 36% written as a decimal?

a. 0.036
b. 0.36
c. 3.6
d. 36
e. NH

6. 4,140 ÷ 36 =

f. 112
g. 113
h. 114
j. 115
k. NH

---

2. $5\frac{1}{2} - 3\frac{3}{4}$ =

f. $1\frac{1}{2}$ j. $2\frac{1}{2}$

g. $1\frac{3}{4}$ k. NH

h. $2\frac{1}{4}$

7. $\frac{3}{8} \times \frac{4}{9}$ =

a. $\frac{1}{6}$ d. $\frac{2}{3}$

b. $\frac{7}{17}$ e. NH

c. $\frac{1}{3}$

---

3. What is $\frac{3}{4}$ written as a percent?

a. 75%
b. 80%
c. 85%
d. 34%
e. NH

8. If n = 7, then 5n + 6 =

f. 89
g. 40
h. 41
j. 42
k. NH

---

4. 329 ÷ 30 is between which numbers?

f. 7 and 8
g. 8 and 9
h. 9 and 10
j. 10 and 11
k. NH

9. What is 0.38 written as a fraction in its lowest terms?

a. $\frac{38}{100}$ d. $\frac{18}{50}$

b. $\frac{38}{50}$ e. NH

c. $\frac{19}{50}$

---

5. If $\frac{n}{6}$ = 9, then n =

a. 45
b. 54
c. 56
d. 63
e. NH

10. 3.6 x 8.01 =

f. 27.836
g. 28.836
h. 26.286
j. 25.236
k. NH

Name _____

Directions

Read each question and choose the correct answer. Mark the space for the answer you have chosen. Mark NH if the answer is not here.

1. Patrick carried the football on two plays. He lost two yards then gained 6. What was his net gain or loss for the two plays?

   a. 3-yard loss
   b. 8-yard gain
   c. 8-yard loss
   d. 3-yard gain
   e. NH

2. An elevator on the seventh floor goes up 3 floors, then down 6 floors. On which floor is the elevator now?

   f. fourth floor
   g. fifth floor
   h. sixth floor
   j. seventh floor
   k. NH

3. Justin had $35 in his savings account. He deposited $25. How much does Justin have in his account now?

   a. $45
   b. $50
   c. $55
   d. $60
   e. NH

4. An airplane was flying at an altitude of 25,000 feet. It descended 10,000 feet. At what altitude is the plane now?

   f. 10,000 feet
   g. 15,000 feet
   h. 20,000 feet
   j. 25,000 feet
   k. NH

5. The actual temperature is 28°F. The wind chill makes it feel 15° colder. What is the temperature with the wind chill?

   a. 43° F
   b. 23° F
   c. 13° F
   d. 11° F
   e. NH

6. Rebecca had $19.82 in her bank. She took out $6.75 to buy a new book. How much money is left in her bank?

   f. $12.07
   g. $12.87
   h. $12.97
   j. $13.07
   k. NH

7. The temperature at noon was 71°F. At 5:00 p.m., the temperature was 59°F. How much did the temperature drop in those five hours?

   a. 12° F
   b. 11° F
   c. 10° F
   d. 9° F
   e. NH

8. Amanda is standing on a hill that is 17 feet above sea level. Her sister is in a valley that is 2 feet below sea level. What is the difference in the two elevations?

   f. 17 feet
   g. 18 feet
   h. 19 feet
   j. 20 feet
   k. NH

Name _____

Directions

Use the coordinate points to answer the questions. Read each question and choose the correct answer. Mark the space for the answer you have chosen. Mark NH if the answer is not here.

Given the following:

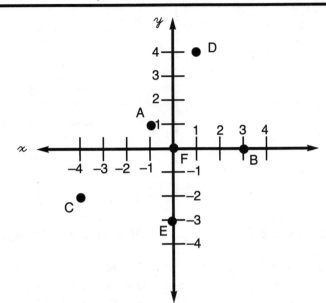

1. What are the coordinates of point A?

    a. (–1, 1)
    b. (1, 1)
    c. (–1, –1)
    d. (1, 1)
    e. NH

2. What are the coordinates of point B?

    f. (0, 3)
    g. (3, 1)
    h. (3, 0)
    j. (1, 3)
    k. NH

3. What are the coordinates of point C?

    a. (–2, –4)
    b. (–4, –2)
    c. (–4, 2)
    d. (2, –4)
    e. NH

4. What are the coordinates of point D?

    f. (1, 4)
    g. (4, 1)
    h. (4, –1)
    j. (4, 4)
    k. NH

5. What are the coordinates of point E?

    a. (–3, 0)
    b. (3, 0)
    c. (0, 3)
    d. (0, –3)
    e. NH

6. What are the coordinates of point F?

    f. (–1, –1)
    g. (1, 1)
    h. (1, 0)
    j. (0, 0)
    k. NH

Directions

Use the coordinate points to answer the questions. Read each question and choose the correct answer. Mark the space for the answer you have chosen. Mark NH if the answer is not here.

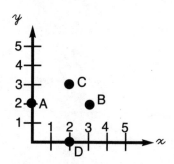

1. At what point does x = 2 and y = 3?

a. **A**
b. **B**
c. **C**
d. **D**
e. **NH**

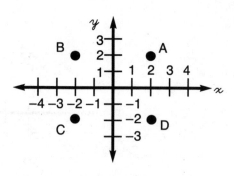

3. At what point does x = −2 and y = −2?

a. **A**
b. **B**
c. **C**
d. **D**
e. **NH**

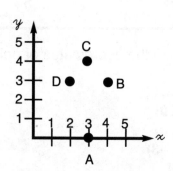

2. At what point does x = 3 and y = 4?

f. **A**
g. **B**
h. **C**
j. **D**
k. **NH**

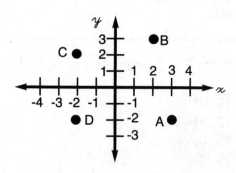

4. What point is located at (−2, 2)?

f. **A**
g. **B**
h. **C**
j. **D**
k. **NH**

Name _____

Directions

Read each question and choose the correct answer. Mark the space for the answer you have chosen. Mark NH if the answer is not here.

---

1. Taylor wanted her mom to buy a box of cereal that cost $4.29. If her mom had a 75¢ coupon for it, how much would the cereal cost?

   a. $5.04
   b. $4.54
   c. $4.45
   d. $3.54
   e. NH

5. Jaime charges $3 per hour for baby-sitting. She sat 3 hours on Friday and 5 hours on Saturday. How much money did she make?

   a. $8
   b. $9
   c. $21
   d. $24
   e. NH

---

2. Johnny bought a magazine for $3.95. If tax was 32¢, how much change did he receive from a $5.00 bill?

   f. 73¢
   g. 69¢
   h. 63¢
   j. 61¢
   k. NH

6. Betsy's hamburger meal cost $3.52 including tax. How much change did she receive from a $10.00 bill?

   f. $6.52
   g. $6.48
   h. $6.42
   j. $7.48
   k. NH

---

3. Kent has $194.14 in his checking account. If he wrote a check for $42.95, what is his new balance in the account?

   a. $152.91
   b. $151.19
   c. $151.91
   d. $115.19
   e. NH

7. The new Peanut-Butter CD costs $14.95. If tax is $1.20, what is the total cost of the CD?

   a. $14.95
   b. $15.25
   c. $15.95
   d. $16.15
   e. NH

---

4. Klean-Kar charges $12 for a wash, $3 for a wax, and $10 for cleaning the inside of the car. If you want all these services, how much will it cost?

   f. $15
   g. $22
   h. $23
   j. $25
   k. NH

8. Brian earns $10 a week for mowing the lawn. How long must he save to buy a new video game for $52.95?

   f. 5 weeks
   g. 6 weeks
   h. 7 weeks
   j. 8 weeks
   k. NH

Name _____

Directions

Read each question and choose the correct answer. Mark the space for the answer you have chosen. Mark NH if the answer is not here.

1. Sippy's Slush-Cones cost 95¢ each. It is 15¢ extra for cream flavors. Bailey wants two strawberry cream cones. How much will they cost?

   a. $2.20
   b. $2.10
   c. $1.95
   d. $1.90
   e. NH

5. Peter paid $12.95 for a new CD and $2.95 for a cassette single. If tax was $1.11, what was the total cost?

   a. $16.95
   b. $16.99
   c. $17.01
   d. $17.02
   e. NH

2. Jim bought a new shirt for $23.85 including tax. How much change did he receive from $30?

   f. $5.15
   g. $6.15
   h. $6.65
   j. $7.15
   k. NH

6. It costs 50¢ to play each arcade game. If Sam played 15 different games, how much money did he spend?

   f. $7.00
   g. $7.50
   h. $8.00
   j. $8.50
   k. NH

3. It costs $6 per person for the movie. It costs $4.50 for a bag of popcorn and a soda. How much will it cost a family of 4 to go to the movie and each get popcorn and soda?

   a. $38
   b. $42
   c. $44
   d. $46
   e. NH

7. On Wednesdays, Happy Burgers are on sale for 39¢ each. How much will 5 Happy Burgers cost on Wednesday?

   a. $1.75
   b. $1.85
   c. $1.95
   d. $2.05
   e. NH

4. A pair of shoes usually costs $52.95. This week they are on sale for $15 off. What is the new price of the shoes?

   f. $67.95
   g. $52.80
   h. $42.95
   j. $37.95
   k. NH

8. In one week, Penny earned $8 for baby-sitting and $5 for walking a neighbor's dog. She also got $6 allowance from her parents. How much money did Penny earn?

   f. $11
   g. $13
   h. $18
   j. $19
   k. NH

Name _____   Skill: Averages

Directions
Read each question and choose the correct answer. Mark the space for the answer you have chosen. Mark NH if the answer is not here.

1. Our puppy gained 15 pounds in 3 months. What was his average weight gain per month?

   a. 3 lb
   b. 4 lb
   c. 5 lb
   d. 6 lb
   e. NH

5. The morning temperatures last week were 36°, 35°, 31°, 30°, 36°, 35°, and 28° F. What was the average temperature for the week?

   a. 30° F
   b. 31° F
   c. 32° F
   d. 33° F
   e. NH

2. On her history tests, Jensen earned scores of 78, 85, and 92. What was her average score for the 3 tests?

   f. 83
   g. 84
   h. 85
   j. 86
   k. NH

6. Jordan had 18 hits in the last 9 games. How many hits does he average per game?

   f. 1
   g. 2
   h. 3
   j. 4
   k. NH

3. Pablo earned $300 one month and $450 the next month. What was the average of his earnings for these two months?

   a. $375
   b. $425
   c. $450
   d. $475
   e. NH

7. Tracy bowled three games. Her scores were 129, 157, and 131. What was her average score per game?

   a. 136
   b. 139
   c. 140
   d. 143
   e. NH

4. Eric grew 6 inches in one year. What was his average growth per month?

   f. 1 inch
   g. $\frac{1}{6}$ inch
   h. $\frac{1}{4}$ inch
   j. $\frac{1}{2}$ inch
   k. NH

8. The boys' basketball team scored 56, 39, 41, 54, and 45 points in its last 5 games. How many points did they average per game?

   f. 45 points
   g. 46 points
   h. 47 points
   j. 48 points
   k. NH

Directions
Read each question and choose the correct answer. Mark the space for the answer you have chosen. Mark NH if the answer is not here.

1. If it takes 2 hours to cook a 3-pound roast, how long will it take to cook a 6-pound roast?

   a. 4 hours
   b. 3 hours
   c. 2 hours
   d. 6 hours
   e. NH

2. Marian frosted 5 cupcakes in 2 minutes. At this rate, how many minutes will it take her to frost 20 cupcakes?

   f. 4 minutes
   g. 6 minutes
   h. 8 minutes
   j. 9 minutes
   k. NH

3. On a map, 1 inch is the same as 5 miles. It is 10 inches between two towns on the map. How many miles are between the two towns?

   a. 2 miles
   b. 25 miles
   c. 50 miles
   d. 100 miles
   e. NH

4. In Ms. Jacob's class, there are 4 girls for every 3 boys. If there are 12 boys, what is the number of girls in the class?

   f. 16 girls
   g. 12 girls
   h. 9 girls
   j. 6 girls
   k. NH

5. Our recipe for sugar cookies calls for 2 cups of sugar for each cup of butter. How much sugar is needed if we use 10 cups of butter?

   a. 10 cups
   b. 3 cups
   c. 5 cups
   d. 20 cups
   e. NH

6. At the car wash fund-raiser, we washed 3 vans for every 2 cars. If we washed a total of 12 cars, how many vans did we wash?

   f. 8 vans
   g. 18 vans
   h. 20 vans
   j. 32 vans
   k. NH

7. Our club made $5 for every 3 boxes of candy we sold. We sold a total of 183 boxes of candy. How much money did we make?

   a. $275
   b. $285
   c. $295
   d. $305
   e. NH

8. Dad mixed paint today. He used 3 pints of white for every 2 gallons of brown. How many gallons of brown did he need if he used 6 pints of white?

   f. 8 gallons
   g. 6 gallons
   h. 4 gallons
   j. 2 gallons
   k. NH

Directions
Read each question and choose the correct answer. Mark the space for the answer you have chosen. Mark NH if the answer is not here.

---

1. A bag contains 3 blue marbles, 5 red marbles, and 4 yellow marbles. What is the probability that a marble chosen at random will be red?

   a. $\frac{5}{12}$ d. $\frac{5}{8}$

   b. $\frac{3}{12}$ e. NH

   c. $\frac{5}{7}$

5. If you roll a six-sided die, what is the probability that you will roll an even number?

   a. $\frac{1}{6}$ d. $\frac{1}{2}$

   b. $\frac{1}{4}$ e. NH

   c. $\frac{1}{3}$

---

2. A bag contains 3 blue marbles, 5 red marbles, and 4 yellow marbles. What is the probability that a marble chosen at random will be blue or red?

   f. $\frac{5}{12}$ j. $\frac{1}{5}$

   g. $\frac{1}{4}$ k. NH

   h. $\frac{2}{3}$

6. What is the probability that a day of the week chosen at random will begin with the letter "S"?

   f. $\frac{2}{7}$ j. $\frac{4}{5}$

   g. $\frac{1}{7}$ k. NH

   h. $\frac{3}{7}$

---

3. A bag contains 3 blue marbles, 5 red marbles, and 4 yellow marbles. What is the probability that a marble chosen at random will <u>not</u> be blue?

   a. $\frac{3}{4}$ d. $\frac{1}{4}$

   b. $\frac{2}{3}$ e. NH

   c. $\frac{1}{2}$

7. What is the probability that a month of the year chosen at random will have 30 or more days?

   a. $\frac{3}{4}$ d. $\frac{12}{12}$

   b. $\frac{5}{12}$ e. NH

   c. $\frac{11}{12}$

---

4. If you roll a six-sided die, what is the probability that you will roll a 2?

   f. $\frac{2}{3}$ j. $\frac{1}{4}$

   g. $\frac{1}{2}$ k. NH

   h. $\frac{1}{6}$

8. If you choose a letter of the alphabet at random, what is the probability that the letter will be a vowel?

   f. $\frac{5}{26}$ j. $\frac{5}{24}$

   g. $\frac{9}{26}$ k. NH

   h. $\frac{11}{24}$

Directions

Read each question and choose the number sentence you can use to find each answer. Mark the space for the number sentence you have chosen. Mark NH if the answer is not here.

1. Which number sentence means, "The sum of five and a number (n)"?

   a. $n - 5$
   b. $5 + n$
   c. $5n$
   d. $\frac{5}{n}$
   e. NH

5. Which number sentence means, "The quotient of a number (n) and nine"?

   a. $n \div 9$
   b. $9n$
   c. $n + 9$
   d. $n - 9$
   e. NH

2. Which number sentence means, "The product of six and a number (n)"?

   f. $6 \div c$
   g. $6 + n$
   h. $6 - n$
   j. $6n$
   k. NH

6. Which number sentence means, "The difference of a number (n) and five"?

   f. $n + 5$
   g. $5n$
   h. $n - 5$
   j. $n \div 5$
   k. NH

3. Which number sentence means, "Seven more than twice a number (n)"?

   a. $7 + 2n$
   b. $14n$
   c. $7 - 2n$
   d. $2n \div 7$
   e. NH

7. Which number sentence means, "Six more than four times a number (n)"?

   a. $4 + 6n$
   b. $4 \times 6n$
   c. $6 + 4n$
   d. $4n - 6$
   e. NH

4. Which number sentence means, "One less than a number (n)"?

   f. $1 - n$
   g. $1n$
   h. $n \div 1$
   j. $n - 1$
   k. NH

8. Which number sentence means, "Twice a number (n)"?

   f. $2 + n$
   g. $2 \div n$
   h. $2n$
   j. $n \div 2$
   k. NH

Name _____

Directions

Read each question and choose the correct answer. Mark the space for the answer you have chosen. Mark NH if the answer is not here.

1. Which number sentence means, "Twice a number (n) and five is eleven"?

   a. $2n \div 5 = 11$
   b. $2n \times 5 = 11$
   c. $2n - 5 = 11$
   d. $2n + 5 = 11$
   e. NH

5. Which number sentence means, "Four times a number (n) minus nine is thirteen"?

   a. $(4n)(9) = 13$
   b. $4n \div 9 = 13$
   c. $4n - 9 = 13$
   d. $4n + 9 = 13$
   e. NH

2. Which number sentence means, "One less than three times a number (n) is six"?

   f. $3n \div 1 = 6$
   g. $3n - 1 = 6$
   h. $3n + 1 = 6$
   j. $3n \times 1 = 6$
   k. NH

6. Which number sentence means, "The difference of three times a number (n) and seven is fourteen"?

   f. $3b + 7 = 14$
   g. $3n \div 7 = 14$
   h. $3n - 7 = 14$
   j. $3n \times 7 = 14$
   k. NH

3. Which number sentence means, "The sum of a number (n) and twelve is eighteen"?

   a. $n - 12 = 18$
   b. $n + 12 = 18$
   c. $n \div 12 = 18$
   d. $n \times 12 = 18$
   e. NH

7. Which number sentence means, "The product of five times a number (n) is fifteen"?

   a. $5n = 15$
   b. $5 \div n = 15$
   c. $5 + n = 15$
   d. $5 - n = 15$
   e. NH

4. Which number sentence means, "The quotient of a number (n) and four is twelve"?

   f. $n \div 4 = 12$
   g. $4n = 12$
   h. $n - 4 = 12$
   j. $n + 4 = 12$
   k. NH

8. Which number sentence means, "Twelve times a number (n) and eight is twenty"?

   f. $(12n)(8) = 20$
   g. $12n \div 8 = 20$
   h. $12n - 8 = 20$
   j. $12n + 8 = 20$
   k. NH

Directions
Read each question and choose the correct answer. Mark the space for the answer you have chosen. Mark NH if the answer is not here.

1. Which angle is <u>not</u> acute?

a.

c.

b.

d.

      e. NH

5. Which of the following could <u>not</u> be the measure of an acute angle?

a. 52°
b. 57°
c. 110°
d. 15°
e. NH

2. Which angle measures 90°?

f.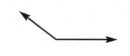

h.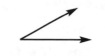

g.

j.

      k. NH

6. What is the measure of an obtuse angle?

f.   **exactly 180°**
g. **less than 90°**
h. **between 90° and 180°**
j.   **more than 180°**
k. **NH**

3. How would you classify this angle?

a. **acute**
b. **obtuse**
c. **right**
d. **straight**
e. **NH**

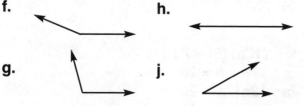

7. How would a 125° angle be classified?

a. **acute**
b. **right**
c. **straight**
d. **obtuse**
e. **NH**

4. Which angle would measure about 170°?

f.

h.

g.

j.

      k. NH

8. Which of the following is <u>not</u> a measure of an obtuse angle?

f.   157°
g. 92°
h. 90°
j.   173°
k. NH

Directions
Read each question and choose the correct answer. Mark the space for the answer you have
chosen. Mark NH if the answer is not here.

---

1. Which angle is not obtuse?

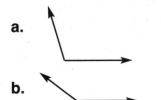

a.

c.

b.

d.

e. NH

---

5. Which of the following could not be the
measure of an obtuse angle?

a. 120°
b. 115°
c. 110°
d. 15°
e. NH

---

2. Which angle measures 180°?

f.

h.

g.

j.

k. NH

---

6. What is the measure of an acute angle?

f. exactly 180°
g. less than 90°
h. between 90° and 180°
j. more than 180°
k. NH

---

3. How would you classify this angle?

a. acute
b. obtuse
c. right
d. straight
e. NH

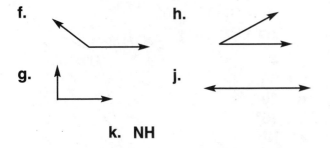

---

7. How would a 65° angle be classified?

a. acute
b. right
c. straight
d. obtuse
e. NH

---

4. Which angle would measure 180°?

f.

h.

g.

j.

k. NH

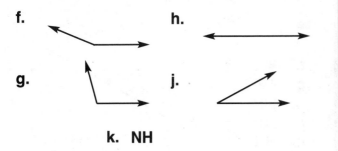

---

8. Which of the following is a measure of an
obtuse angle?

f. 157°
g. 92°
h. 90°
j. 17°
k. NH

Directions

Read each question and choose the correct answer. Mark the space for the answer you have chosen. Mark NH if the answer is not here.

---

1. What is the measure of angle BCA given the following:

   Angle ABC = 90°
   Angle BAC = 30°

   a. 30°
   b. 60°
   c. 70°
   d. 90°
   e. NH

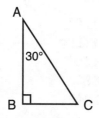

5. What is the measure of angle RTS given the following:

   Angle SRT = 50°
   Angle TSR = 40°

   a. 90°
   b. 80°
   c. 70°
   d. 60°
   e. NH

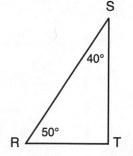

---

2. If $\overline{AB}$ = 4 inches, what is the measurement of $\overline{BC}$?

   f. 1 inch
   g. 2 inches
   h. 3 inches
   j. 4 inches
   k. NH

6. What is the measure of angle PQR given the following:

   Angle RPQ = 130°
   Angle PRQ = 30°

   f. 40°
   g. 30°
   h. 20°
   j. 10°
   k. NH

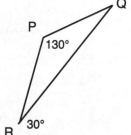

---

3. If angles N, M, and O have equal measures, what is the measure of each angle?

   a. 20°
   b. 30°
   c. 40°
   d. 60°
   e. NH

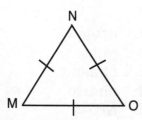

7. What is the measure of angle WUV given the following:

   Angle UVW = 35°
   Angle VWU = 88°

   a. 43°
   b. 47°
   c. 53°
   d. 57°
   e. NH

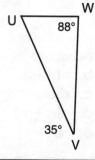

---

4. What is the measure of angle Y given the following:

   Angle YXZ = 50°
   Angle YZX = 50°

   f. 80°
   g. 50°
   h. 60°
   j. 70°
   k. NH

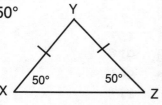

8. Angles M and T are equal. What is the measure of angle M?

   f. 60°
   g. 40°
   h. 30°
   j. 20°
   k. NH

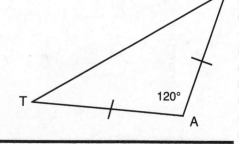

Directions
Read each question and choose the correct answer. Mark the space for the answer you have chosen. Mark NH if the answer is not here.

---

1. What of the following names a diameter of circle B?

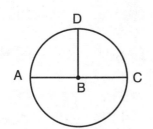

a. $\overline{BD}$
b. $\overline{AC}$
c. $\overline{BA}$
d. $\overline{BC}$
e. NH

5. Which of the following is <u>not</u> a radius of circle C?

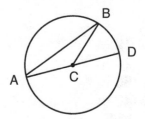

a. $\overline{AC}$
b. $\overline{AB}$
c. $\overline{BC}$
d. $\overline{CD}$
e. NH

---

2. If the diameter of this circle is 8 inches, what is the radius of this circle?

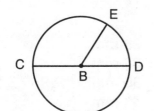

f. 4 in
g. 8 in
h. 12 in
j. 16 in
k. NH

6. If $\overline{BC}$ is 20 inches long, what is the measure of $\overline{AC}$?

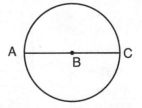

f. 10 in
g. 15 in
h. 20 in
j. 40 in
k. NH

---

3. If $\overline{LN}$ measures 10 millimeters, what is the measure of $\overline{PM}$?

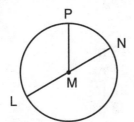

a. 20 mm
b. 15 mm
c. 10 mm
d. 5 mm
e. NH

7. If $\overline{BM}$ measures 14 centimeters, what is the measure of $\overline{KM}$?

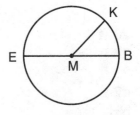

a. 28 cm
b. 14 cm
c. 7 cm
d. 3.5 cm
e. NH

---

4. Which is a chord of this circle?

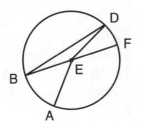

f. $\overline{EF}$
g. $\overline{BD}$
h. $\overline{CE}$
j. $\overline{AE}$
k. NH

8. If the radius of this circle is 12 meters, what is the measure of the diameter?

f. 6 m
g. 18 m
h. 24 m
j. 48 m
k. NH

Directions
Read each question and choose the correct answer. Mark the space for the answer you have
chosen. Mark NH if the answer is not here.

1. What is the perimeter of this square?

a. **12 in**
b. **24 in**
c. **36 in**
d. **48 in**
e. **NH**

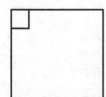
6 in

5. What is the perimeter of this parallelogram?

a. **9 in**
b. **12 in**
c. **15 in**
d. **18 in**
e. **NH**

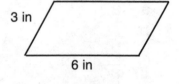
3 in
6 in

2. What is the perimeter of this rectangle?

f. **13 m**
g. **18 m**
h. **26 m**
j. **36 m**
k. **NH**

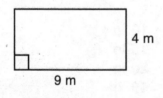
4 m
9 m

6. What is the perimeter of this triangle?

f. **25 mm**
g. **30 mm**
h. **35 mm**
j. **40 mm**
k. **NH**

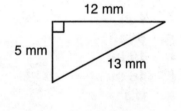
12 mm
5 mm
13 mm

3. What is the perimeter of this trapezoid?

a. **26 ft**
b. **28 ft**
c. **32 ft**
d. **42 ft**
e. **NH**

6 ft
8 ft   7 ft
11 ft

7. What is the perimeter of this square?

a. **19 m**
b. **21 m**
c. **27 m**
d. **38 m**
e. **NH**

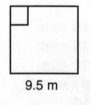
9.5 m

4. What is the perimeter of this triangle?

f. **15 cm**
g. **18 cm**
h. **19 cm**
j. **21 cm**
k. **NH**

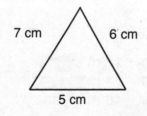
7 cm   6 cm
5 cm

8. What is the perimeter of this pentagon?

f. **20 cm**
g. **25 cm**
h. **30 cm**
j. **35 cm**
k. **NH**

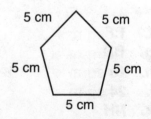
5 cm   5 cm
5 cm   5 cm
5 cm

Name _____   Skill: Area

Directions
Read each question and choose the correct answer. Mark the space for the answer you have chosen. Mark NH if the answer is not here.

---

1. What is the area of this rectangle in square centimeters?

8 cm

a. **48 sq cm**
b. **28 sq cm**   6 cm
c. **24 sq cm**
d. **14 sq cm**
e. **NH**

---

5. What is the area of the triangle in square millimeters?

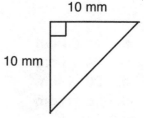

a. **30 sq mm**
b. **50 sq mm**   10 mm
c. **100 sq mm**
d. **150 sq mm**
e. **NH**

---

2. What is the area of the triangle in square inches?

f. **15 sq in**   10 in
g. **16 sq in**
h. **30 sq in**
j. **60 sq in**
k. **NH**   6 in

---

6. What is the area of this parallelogram in square inches?

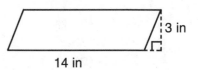

f. **17 sq in**
g. **21 sq in**
h. **28 sq in**
j. **42 sq in**
k. **NH**   14 in

---

3. What is the area of this square in square feet?

a. **18 sq ft**
b. **36 sq ft**
c. **98 sq ft**
d. **81 sq ft**
e. **NH**   9 ft

---

7. What is the area of a rectangle with a length of 6 feet and a width of 7 feet?

a. **42 sq ft**
b. **40 sq ft**
c. **34 sq ft**
d. **28 sq ft**
e. **NH**

---

4. What is the area of this parallelogram in square meters?

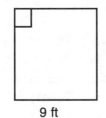

f. **12 sq m**   2 m
g. **18 sq m**
h. **20 sq m**   10 m
j. **24 sq m**
k. **NH**

---

8. What is the area of a square with a side of 11 meters?

f. **22 sq m**
g. **110 sq m**
h. **121 sq m**
j. **132 sq m**
k. **NH**

Name _____

Directions
Read each question and choose the correct answer. Mark the space for the answer you have chosen. Mark NH if the answer is not here.

---

1. What is the volume of this box in cubic centimeters?

a. **48 cu cm**
b. **24 cu cm**
c. **20 cu cm**
d. **16 cu cm**
e. **NH**

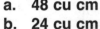
12 cm
2 cm
2 cm

---

5. What is the volume of this box in cubic meters?

a. **96 cu m**
b. **45 cu m**
c. **30 cu m**
d. **15 cu m**
e. **NH**

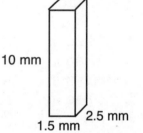
4 m
3 m
8 m

---

2. What is the volume of this cube in cubic inches?

f. **15 cu in**
g. **125 cu in**
h. **25 cu in**
j. **50 cu in**
k. **NH**

5 in
5 in
5 in

---

6. What is the volume of this prism in cubic millimeters?

f. **37.5 cu mm**
g. **28.5 cu mm**
h. **18 cu mm**
j. **14 cu mm**
k. **NH**

10 mm
2.5 mm
1.5 mm

---

3. What is the volume of this box in cubic feet?

a. **30 cu ft**
b. **21 cu ft**
c. **18 cu ft**
d. **15 cu ft**
e. **NH**

2 ft
5 ft
3 ft

---

7. What is the volume of this box in cubic meters?

a. **186 cu m**
b. **168 cu m**
c. **148 cu m**
d. **96 cu m**
e. **NH**

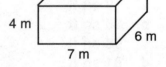
4 m
6 m
7 m

---

4. What is the volume of this cube in cubic centimeters?

f. **9 cu cm**
g. **27 cu cm**
h. **81 cu cm**
j. **243 cu cm**
k. **NH**

3 cm

---

8. What is the volume of this cube in cubic feet?

f. **16 cu ft**
g. **64 cu ft**
h. **24 cu ft**
j. **48 cu ft**
k. **NH**

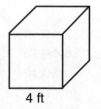
4 ft

Name _____

Directions
Use the frequency table to answer the questions. Read each question and choose the correct answer. Mark the space for the answer you have chosen. Mark NH if the answer is not here.

The following table shows the number of students earning different scores on the math exam. Use the table to answer the questions.

MATH EXAM SCORES

Score	Frequency
100	4
95	7
90	3
85	5
80	8
75	2
70	2
65	1
60	0
55	0

3. How many students scored below 72 on the exam?

   a. 2
   b. 3
   c. 4
   d. 5
   e. NH

4. If 69 is the lowest passing score, how many students passed this exam?

   f. 32
   g. 31
   h. 30
   j. 29
   k. NH

5. How many students scored below 85 on this exam?

   a. 16
   b. 15
   c. 14
   d. 13
   e. NH

1. How many students took the exam?

   a. 35
   b. 34
   c. 33
   d. 32
   e. NH

6. What was the grade level of the students taking this exam?

   f. 5th grade
   g. 6th grade
   h. 7th grade
   j. 8th grade
   k. NH

2. How many students scored 95 or higher?

   f. 11
   g. 10
   h. 9
   j. 8
   k. NH

7. How many students passed the test with a score of 80 or higher?

   a. 25
   b. 27
   c. 29
   d. 31
   e. NH

Directions

Use the frequency table to answer the questions. Read each question and choose the correct answer. Mark the space for the answer you have chosen. Mark NH if the answer is not here.

Jenny's class took a survey to find favorite types of fast food. The results are summarized in the following table. Use the table to answer the questions.

FAVORITE FOODS

Types	Frequency
Hamburgers	87
Pizza	112
Hot Dogs	29
Chicken	58
Tacos	78
Other	15

1. How many people participated in this survey?

   a. 379
   b. 381
   c. 383
   d. 385
   e. NH

2. What type of fast food was chosen as a favorite most often?

   f. **tacos**
   g. **hamburgers**
   h. **pizza**
   j. **chicken**
   k. **NH**

3. How many more people chose tacos than hot dogs?

   a. 62
   b. 58
   c. 49
   d. 46
   e. NH

4. How many people chose chicken or hamburgers as their favorite fast food?

   f. 141
   g. 143
   h. 145
   j. 147
   k. NH

5. What type of fast food had the lowest frequency according to this survey?

   a. **hamburgers**
   b. **hot dogs**
   c. **chicken**
   d. **other**
   e. **NH**

6. How many more people chose pizza than hot dogs and chicken?

   f. 25
   g. 27
   h. 29
   j. 31
   k. NH

7. How many people did not choose pizza as their favorite fast food?

   a. 267
   b. 269
   c. 271
   d. 275
   e. NH

Name _____ Skill: Fraction Word Problems

Directions
Read each question and choose the correct answer. Mark the space for the answer you have chosen. Mark NH if the answer is not here.

---

1. Kaitlin has saved $30 toward a $40 sweater she wants to buy. What fraction of the selling price has she saved?

   a. $\frac{1}{2}$ d. $\frac{4}{5}$

   b. $\frac{2}{3}$ e. NH

   c. $\frac{3}{4}$

5. Hallie has saved $50 toward the $75 skates she wants. What fraction of the skates' cost has she saved?

   a. $\frac{1}{3}$ d. $\frac{3}{4}$

   b. $\frac{2}{3}$ e. NH

   c. $\frac{1}{2}$

---

2. Eric earned $15 cutting grass. He wants to buy a $50 video game. What fraction of the selling price did he earn cutting grass?

   f. $\frac{1}{10}$ j. $\frac{2}{5}$

   g. $\frac{1}{5}$ k. NH

   h. $\frac{3}{10}$

6. Jim has saved $90 toward the new $120 bicycle he wants. What fraction of the price does he have?

   f. $\frac{12}{3}$ j. $\frac{5}{6}$

   g. $\frac{3}{4}$ k. NH

   h. $\frac{4}{5}$

---

3. Matthew has saved $80 toward a new $120 catcher's mitt. What fraction of the price has Matthew saved?

   a. $\frac{1}{2}$ d. $\frac{5}{6}$

   b. $\frac{2}{3}$ e. NH

   c. $\frac{3}{4}$

7. Catherine has $5 for the new movie she wants to buy. If the movie costs $20, what fraction of the movie's cost does she have?

   a. $\frac{1}{4}$ d. $\frac{1}{5}$

   b. $\frac{1}{2}$ e. NH

   c. $\frac{1}{3}$

---

4. Kara has $25 of the $35 she needs for the concert ticket. What fraction of the ticket's cost does she have?

   f. $\frac{5}{7}$ j. $\frac{4}{5}$

   g. $\frac{5}{6}$ k. NH

   h. $\frac{3}{4}$

8. Jesse has $60 of the $70 he needs for new soccer shoes. What fraction of the shoes' cost does he have?

   f. $\frac{5}{6}$ j. $\frac{5}{8}$

   g. $\frac{6}{7}$ k. NH

   h. $\frac{5}{7}$

Directions
Read each question and choose the correct answer. Mark the space for the answer you have chosen. Mark NH if the answer is not here.

1. If Nicole paid $5 for $\frac{1}{2}$ of a pound of cashews, what is the cost of one pound of cashews?

   a. $10.00
   b. $7.50
   c. $5.00
   d. $2.50
   e. NH

5. Salvadore bought $\frac{1}{2}$ of a gallon of gas for 70¢. How much is the gas per gallon?

   a. $1.40
   b. $1.60
   c. $1.80
   d. $2.10
   e. NH

2. Tina bought $\frac{1}{4}$ of a pound of sunflower seeds for 75¢. At that rate, how much does a pound of sunflower seeds cost?

   f. $1.50
   g. $1.75
   h. $3.00
   j. $3.50
   k. NH

6. Lori bought $\frac{1}{2}$ of a pound of chocolate stars for $2.10. How much do the chocolate stars cost per pound?

   f. $1.05
   g. $4.20
   h. $5.25
   j. $6.50
   k. NH

3. Josh bought $\frac{1}{3}$ of a pound of grapes for 60¢. At that rate, how much does a pound of grapes cost?

   a. 20¢
   b. $1.20
   c. $1.80
   d. $2.40
   e. NH

7. Caroline bought $\frac{1}{3}$ of a pound of gummy candy for $1.20. What is the cost of gummy candy per pound?

   a. 40¢
   b. $2.40
   c. $3.60
   d. $4.80
   e. NH

4. Lindsey bought $\frac{1}{5}$ of a pound of sugar cookies for 30¢. How much would a pound of sugar cookies cost?

   f. 60¢
   g. $1.20
   h. $1.50
   j. $1.80
   k. NH

8. Brandon bought $\frac{1}{2}$ of a liter of soda for 75¢. How much does that soda cost per liter?

   f. $1.50
   g. $1.75
   h. $2.25
   j. $2.50
   k. NH

Directions

Read each question and choose the correct answer. Mark the space for the answer you have chosen. Mark NH if the answer is not here.

---

1. 60% of the sixth-grade class voted for Joey for class president. There are 210 students in the sixth grade. How many students voted for Joey?

   a. 126
   b. 128
   c. 130
   d. 132
   e. NH

5. Kara bought a poster. She spent 40% of the $20 she earned baby-sitting. How much did she spend on the poster?

   a. $6.00
   b. $8.00
   c. $10.00
   d. $12.00
   e. NH

---

2. 80% of the 30 girls at the party wanted to dance with Josh. How many girls wanted to dance with Josh?

   f. 20
   g. 21
   h. 24
   j. 27
   k. NH

6. 25% of the proceeds from a fund-raiser were donated to the school to buy some trees. If $250 was raised, how much did the school get for new trees?

   f. $62.50
   g. $65.00
   h. $70.50
   j. $75.00
   k. NH

---

3. Last year, Carey earned 10% interest on $150 in her savings account. How much interest did she earn last year?

   a. $10.00
   b. $15.00
   c. $20.00
   d. $25.00
   e. NH

7. In a recent election, Mary received 60% of the 1,200 votes that were cast. How many votes did Mary receive?

   a. 680
   b. 720
   c. 760
   d. 820
   e. NH

---

4. Mrs. Chavez's class donated 50% of the $60 they earned from the paper drive to the Band Boosters Club. How much money did they donate to the Band Boosters?

   f. $11.00
   g. $15.00
   h. $25.00
   j. $30.00
   k. NH

8. The sixth-grade science class found that 20% of the students in the school had blue eyes. If there were 800 students in the school, how many had blue eyes?

   f. 130
   g. 140
   h. 150
   j. 160
   k. NH

Name _____

Directions
Read each question and choose the correct answer. Mark the space for the answer you have chosen. Mark NH if the answer is not here.

1. Tanya worked from 5:15 p.m. until 8:45 p.m. How many hours did Tanya work?

    a. 3 hours
    b. 3 hours and 30 minutes
    c. 4 hours
    d. 4 hours and 30 minutes
    e. NH

5. Happy Burgers serves breakfast from 7:00 a.m. until 10:30 a.m. How long do they serve breakfast?

    a. 3 hours and 30 minutes
    b. 3 hours
    c. 2 hours and 30 minutes
    d. 2 hours
    e. NH

2. The plane left Miami at 8:30 a.m. and arrived in Boston at 12:30 p.m. How many hours was the flight from Miami to Boston?

    f. 2 hours
    g. 3 hours
    h. 4 hours
    j. 5 hours
    k. NH

6. We shopped at the mall from 11:00 a.m. until 3:45 p.m. How long were we at the mall?

    f. 3 hours and 45 minutes
    g. 4 hours
    h. 4 hours and 15 minutes
    j. 4 hours and 45 minutes
    k. NH

3. Mom started the soup at 2:30 p.m. and let it cook until we ate dinner at 7:00 p.m. How many hours did the soup cook?

    a. 4 hours and 30 minutes
    b. 4 hours
    c. 3 hours and 30 minutes
    d. 3 hours
    e. NH

7. The store opens daily at 8:00 a.m. and closes at 9:00 p.m. How many hours is the store open daily?

    a. 10 hours
    b. 11 hours
    c. 12 hours
    d. 13 hours
    e. NH

4. The movie started at 7:30 p.m. If the show lasted for $2\frac{1}{2}$ hours, what time did it end?

    f. 10:00 p.m.
    g. 9:30 p.m.
    h. 11:00 p.m.
    j. 9:00 p.m.
    k. NH

8. The hockey game started at 7:00 p.m. and lasted for 2 hours and 45 minutes. At what time did the game end?

    f. 9:15 p.m.
    g. 9:30 p.m.
    h. 9:45 p.m.
    j. 10:00 p.m.
    k. NH

Name _____

Directions
Read each question and choose the correct answer. Mark the space for the answer you have chosen. Mark NH if the answer is not here.

1. 10 centimeters =

   a. 1 meter
   b. 100 millimeters
   c. 1 millimeter
   d. 1,000 meters
   e. NH

5. 250 millimeters =

   a. 25 decimeters
   b. 25 centimeters
   c. 25 meters
   d. 25 kilometers
   e. NH

2. 1,000 milligrams =

   f. 1,000 grams
   g. 100 grams
   h. 10 grams
   j. 1 gram
   k. NH

6. 800 grams =

   f. 8,000 kilograms
   g. 80 kilograms
   h. 8 kilograms
   j. 0.8 kilograms
   k. NH

3. 1 meter =

   a. 10,000 millimeters
   b. 1,000 millimeters
   c. 100 millimeters
   d. 10 millimeters
   e. NH

7. 4 kilometers =

   a. 40,000 meters
   b. 4,000 meters
   c. 400 meters
   d. 40 meters
   e. NH

4. 20 grams =

   f. 200,000 milligrams
   g. 20,000 milligrams
   h. 2,000 milligrams
   j. 200 milligrams
   k. NH

8. 1 liter =

   f. 10 milliliters
   g. 100 milliliters
   h. 1,000 milliliters
   j. 10,000 milliliters
   k. NH

Name _____

Directions

Use the graph to answer the questions. Read each question and choose the correct answer. Mark the space for the answer you have chosen. Mark NH if the answer is not here.

This graph shows the results of a survey of sixth graders naming their favorite ice cream flavors.

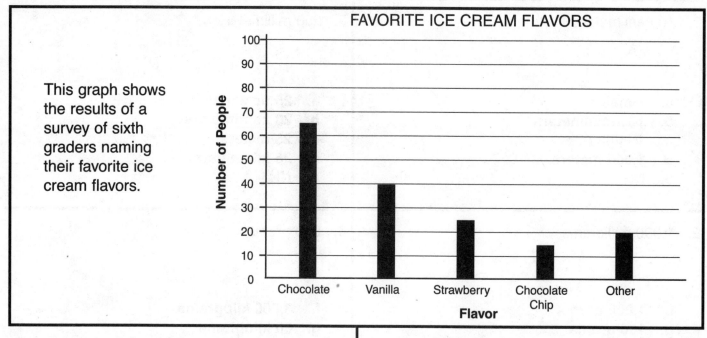

1. How many students participated in the survey?

a. 165
b. 160
c. 155
d. 150
e. NH

2. What flavor was the most popular in this survey?

f. vanilla
g. strawberry
h. chocolate
j. chocolate chip
k. NH

3. How many students like strawberry ice cream the best?

a. 25
b. 30
c. 35
d. 40
e. NH

4. How many students prefer a flavor other than vanilla?

f. 110
g. 115
h. 120
j. 125
k. NH

5. How many students prefer chocolate or chocolate chip ice cream?

a. 80
b. 75
c. 70
d. 65
e. NH

6. What were the flavors people named in the group labeled "other"?

f. **maple pecan and mint**
g. **mint and tutti-frutti**
h. **peach and black cherry**
j. **french twist and blueberry**
k. **NH**

Name _____   Skill: Line Graph

Directions
Use the graph to answer the questions. Read each question and choose the correct answer. Mark the space for the answer you have chosen. Mark NH if the answer is not here.

This graph shows the number of cable subscribers in the city of New Times.

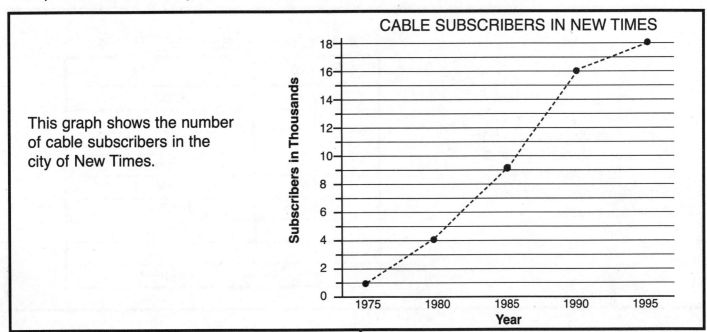

CABLE SUBSCRIBERS IN NEW TIMES

1. About how many cable subscribers were in New Times in 1985?

   a. 6,000
   b. 7,000
   c. 8,000
   d. 9,000
   e. NH

2. About how many more people subscribed to cable between 1985 and 1990?

   f. 5,000
   g. 6,000
   h. 7,000
   j. 9,000
   k. NH

3. Between what years was the greatest increase in cable subscribers?

   a. 1975–1980
   b. 1980–1985
   c. 1985–1990
   d. 1990–1995
   e. NH

4. How many more people subscribed to cable in 1990 than in 1975?

   f. 12,000
   g. 13,000
   h. 14,000
   j. 15,000
   k. NH

5. How many cable subscribers were there in 1995?

   a. 18,000
   b. 16,000
   c. 17,000
   d. 2,000
   e. NH

6. The smallest increase in cable subscribers occurred between what years?

   f. 1975–1980
   g. 1990–1995
   h. 1980–1985
   j. 1985–1990
   k. NH

Name _____   Skill: Mean, Median, Mode, Range

Directions
Use the table to answer the questions. Read each question and choose the correct answer. Mark the space for the answer you have chosen. Mark NH if the answer is not here.

Mrs. Greene keeps a chart of the number of papers each student has completed in her class. Each student needs a total of 10 papers by the end of the semester. This chart shows the results.

NUMBER OF PAPERS PER STUDENT

Papers	Student
8	Abbey
9	Al
6	Fran
8	Bob
6	Kyle
8	Keshia
5	Quincy
8	Tasha
5	Vasquez

1. What is the mean (average) number of papers turned in?

   a. 6
   b. 7
   c. 8
   d. 9
   e. NH

2. What is the mode (most common score) of papers turned in?

   f. 5
   g. 6
   h. 7
   j. 8
   k. NH

3. What is the median (middle) number of papers turned in?

   a. 9
   b. 8
   c. 7
   d. 6
   e. NH

4. What is the range (difference between high and low scores) of papers turned in?

   f. 2
   g. 3
   h. 4
   j. 5
   k. NH

5. What is the average of the four lowest numbers of papers turned in?

   a. 5.5
   b. 5
   c. 4.5
   d. 4
   e. NH

6. If you take out one of the highest and lowest scores on the chart, what is the average?

   f. 4
   g. 5
   h. 6
   j. 7
   k. NH

Directions

Read each question and choose the correct answer. Mark the space for the answer you have chosen.
Mark NH if the answer is not here.

1. The space shuttle was flying 30 miles above the earth. It descended 18 miles. At what altitude is the shuttle flying now?

   a. 48 mi
   b. 38 mi
   c. 12 mi
   d. 21 mi
   e. NH

6. Which angle is acute?

   f. h.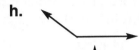

   g. j.

   k. NH

2. What is the measure of angle YXZ given:

   Angle XYZ = 63°
   Angle YZX = 40°

   f. 12°
   g. 13°
   h. 14°
   j. 15°
   k. NH

   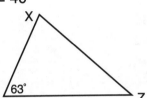

7. There are 7 birds in the pet shop for every 4 fish. If there are 28 fish, how many birds are there in the store?

   a. 32 birds
   b. 32 birds
   c. 16 birds
   d. 9 birds
   e. NH

3. The school lunch costs $1.25 per day. How much would it cost to eat lunch for 42 days?

   a. $53.25
   b. $52.50
   c. $52.75
   d. $53.75
   e. NH

8. What is the volume of a box with a height of 12 centimeters, a width of 4 centimeters, and a length of 6 centimeters?

   f. 22 cm³
   g. 288 cm³
   h. 36 cm³
   j. 45 cm³
   k. NH

4. Which number sentence means the following: The sum of a number and 7 is 33?

   f. n x 7 = 33
   g. n − 7 = 33
   h. n ÷ 7 = 33
   j. n + 7 = 33
   k. NH

9. If you throw a six-sided die, what is the probability that you will roll an even number?

   a. $\frac{3}{6}$ d. $\frac{6}{6}$

   b. $\frac{2}{6}$ e. NH

   c. $\frac{4}{6}$

5. What time is 7 hours and 30 minutes before 9:00 p.m.?

   a. 1:25 p.m.
   b. 1:30 p.m.
   c. 2:25 p.m.
   d. 2:30 p.m.
   e. NH

10. A baseball sells for $9.00. If Steve has $3.60, what fraction of the selling price does he have?

   f. $\frac{1}{2}$ j. $\frac{2}{5}$

   g. $\frac{1}{4}$ k. NH

   h. $\frac{2}{3}$

Name _____

Directions

Read each question and choose the correct answer. Mark the space for the answer you have chosen.
Mark NH if the answer is not here.

---

1. In our club there are 6 girls for every 2 boys. If there are 14 boys, how many girls are there in the club?

   a. 31 girls
   b. 25 girls
   c. 42 girls
   d. 24 girls
   e. NH

6. 12.201 x 6.3507 is closest to which number?

   f. 70
   g. 80
   h. 60
   j. 90
   k. NH

---

2. A car costs $9,000. If Jane has $3,000.00, what fraction of the car's cost does she have?

   f. $\frac{1}{3}$ **j.** $\frac{3}{5}$

   g. $\frac{1}{5}$ **k. NH**

   h. $\frac{2}{3}$

7. Jack had $321 in savings. This week he deposited $31, withdrew $43, and deposited $98 more. What is his balance?

   a. $488
   b. $493
   c. $407
   d. $387
   e. NH

---

3. Which fraction is in its simplest form?

   a. $\frac{5}{6}$ **d.** $\frac{2}{18}$

   b. $\frac{2}{20}$ **e. NH**

   c. $\frac{10}{12}$

8. What is the area of this triangle?

   f. 31 m²
   g. 24 m²
   h. 37 m²
   j. 28 m²
   k. NH

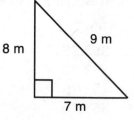

---

4. Which angle would measure 65°?

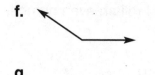

   k. NH

9. Which means the following?
9 times the sum of a number and 8 is 81?

   a. 9n + 8 = 81
   b. 9 (n + 8) = 81
   c. 9n – 8 = 81
   d. 9n x 8 = 81
   e. NH

---

5. $1\frac{1}{3} + 2\frac{2}{4} =$

   a. $3\frac{5}{6}$ **d.** $4\frac{3}{4}$

   b. 2   **e. NH**

   c. $3\frac{2}{3}$

10. 9 meters =

   f. 9,000 kilometers
   g. 90 centimeters
   h. 900 millimeters
   j. 900 centimeters
   k. NH

Name _____

Directions

Read each question and choose the correct answer. Mark the space for the answer you have chosen.
Mark NH if the answer is not here.

---

1. 800 millimeters =

   a. 8 kilometers
   b. 8000 meters
   c. 8000 centimeters
   d. 80 centimeters
   e. NH

6. The weight of four show dogs were 85, 60, 55, and 80 pounds. What was the average weight of the four dogs?

   f. 71 lb
   g. 70 lb
   h. 73 lb
   j. 72 lb
   k. NH

---

2. Which of the following is a radius of the circle?

   f. $\overline{OB}$
   g. $\overline{CB}$
   h. $\overline{AB}$
   j. $\overset{\frown}{AC}$
   k. NH

7. Nancy bought four 21¢ pieces of candy. How much change will she receive from a dollar? (There was no tax.)

   a. $1.00 + (4 \times .21) = n$
   b. $1.00 \div (4 \times .21) = n$
   c. $1.00 \times (4 \times .21) = n$
   d. $1.00 - (4 \times .21) = n$
   e. NH

---

3. Jill bought $\frac{1}{3}$ of a pound of apples for 40¢. What is the cost per pound?

   a. 25¢
   b. 70¢
   c. $1.20
   d. $1.25
   e. NH

8. Jane received 30% of the $95.00 prize money. How much money did Jane receive?

   f. $25.80
   g. $28.50
   h. $28.80
   j. $28.85
   k. NH

---

4. Mike bought eight fishing hooks for $1.29 each. There was no tax. How much did he spend?

   f. $11.31
   g. $12.70
   h. $10.32
   j. $9.87
   k. NH

9. It is now 3:15 a.m. What time will it be in 3 hours and 15 minutes?

   a. 6:05 PM
   b. 6:30 PM
   c. 6:15 PM
   d. 6:25 PM
   e. NH

---

5. Lela spent 70% of her $600 in savings. How much money did Lela spend?

   a. $300
   b. $360
   c. $420
   d. $480
   e. NH

10. Which angle is obtuse?

   f. 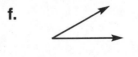   h.

   g.   j.

   k. NH

Name _____

Directions

Read each question and choose the correct answer. Mark the space for the answer you have chosen. Mark NH if the answer is not here.

---

1. 4.205 x 4.545 is closest to which number?

   a. 16
   b. 17
   c. 18
   d. 19
   e. NH

6. What is the perimeter of this triangle?

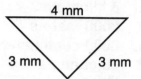

   f. 12 mm
   g. 10 mm
   h. 9 mm
   j. 8 mm
   k. NH

---

2. If Jimmy paid 75¢ for $\frac{1}{2}$ of a pound of sugar, what was the cost for a pound?

   f. $1.05
   g. $1.50
   h. $2.50
   j. $1.00
   k. NH

7. Which of the following is the measure of an obtuse angle?

   a. 89°
   b. 87°
   c. 88°
   d. 91°
   e. NH

---

3. What is the perimeter of this equilateral triangle?

   a. 55 mm
   b. 65 mm
   c. 45 mm
   d. 60 mm
   e. NH

15 mm

8. If you flip a coin, what is the probability that you will get "heads"?

   f. $\frac{3}{5}$ j. $\frac{1}{2}$

   g. $\frac{2}{3}$ k. NH

   h. $\frac{0}{2}$

---

4. Mike bought 3 pieces of candy for 19¢ each. How much change did he receive from a dollar? (There was no tax.)

   f. (19 x 1.00) = n
   g. 1.00 − (3 x .19) = n
   h. 1.00 + 3 + .19 = n
   j. 1.00 − (3 ÷ .19) = n
   k. NH

9. What is the volume of this rectangular prism?

   a. 22 m³
   b. 21 m³
   c. 272.25 m³
   d. 225.7 m³
   e. NH

5.5 m   4.5 m   11 m

---

5. What is the volume of a cube with a side of 3 feet?

   a. 12 ft³
   b. 19 ft³
   c. 81 ft³
   d. 144 ft³
   e. NH

10. If $\overline{AB}$ measures 18 inches, what is the measure of $\overline{DC}$?

   f. 7 in
   g. 8 in
   h. 9 in
   j. 18 in
   k. NH

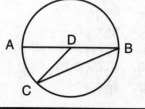

Name _____

Directions

Read each question and choose the correct answer. Mark the space for the answer you have chosen.
Mark NH if the answer is not here.

---

1. Jenny earns $15 a night for baby-sitting.
How many nights must she baby-sit in order
to buy a new dress for $68.50?

   a. **4 nights**
   b. **5 nights**
   c. **6 nights**
   d. **7 nights**
   e. **NH**

6. Which angle is <u>not</u> acute?

   f.   h.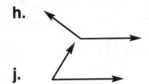

   g.   j.

             k. NH

---

2. Steve bought a new basketball for $32.98
including tax. How much change did he
receive from $40?

   f. **$7.00**
   g. **$7.02**
   h. **$7.50**
   j. **$7.01**
   k. **NH**

7. There are 8 red balls for every 2 green balls.
If there are 10 green balls, how many red
balls are there?

   a. **36 red balls**
   b. **40 red balls**
   c. **12 red balls**
   d. **18 red balls**
   e. **NH**

---

3. Cindy paid $3.50 for a pound of coffee,
$4.50 for a magazine, and $1.20 for a candy
bar. If tax was 88¢, what was the total cost?

   a. **$10.07**
   b. **$10.50**
   c. **$10.08**
   d. **$9.20**
   e. **NH**

8. What is the volume of a box with a height of
11 centimeters, a width of 5 centimeters,
and a length of 7 centimeters?

   f. **23 cm³**
   g. **82 cm³**
   h. **385 cm³**
   j. **405 cm³**
   k. **NH**

---

4. Which number sentence means:
The sum of a number and 9 is 23?

   f. **n x 9 = 23**
   g. **n − 9 = 23**
   h. **n ÷ 9 = 23**
   j. **n + 9 = 23**
   k. **NH**

9. Jason had 20 hits in the last 10 games. How
many hits did he average per game?

   a. **1**
   b. **2**
   c. **3**
   d. **4**
   e. **NH**

---

5. What time is 5 hours and 20 minutes before
10:00 p.m.?

   a. **4:25 p.m.**
   b. **4:40 p.m.**
   c. **4:35 p.m.**
   d. **4:20 p.m.**
   e. **NH**

10. A baseball sells for $8.00. If Steve has $4.00,
what fraction of the selling price does he have?

   f. $\frac{1}{2}$ j. $\frac{2}{5}$

   g. $\frac{1}{4}$ k. **NH**

   h. $\frac{2}{3}$

Name _____

Directions

Read each question and choose the correct answer. Mark the space for the answer you have chosen.
Mark NH if the answer is not here.

1. The cheerleaders made $4 for every 3 boxes of cards they sold. They sold 150 boxes of cards. How much money did they make?

   a. $110
   b. $200
   c. $175
   d. $122
   e. NH

2. A bicycle costs $155. If Lea has $31, what fraction of the car's cost does she have?

   f. $\frac{1}{3}$ j. $\frac{3}{5}$

   g. $\frac{1}{5}$ k. NH

   h. $\frac{2}{3}$

3. Which fraction is in its simplest form?

   a. $\frac{4}{7}$ d. $\frac{3}{18}$

   b. $\frac{3}{33}$ e. NH

   c. $\frac{12}{15}$

4. Which angle would measure closest to 18°?

   f.   h.

   g.   j.

          k. NH

5. $2\frac{1}{5} + 3\frac{2}{10} =$

   a. $5\frac{5}{6}$ d. $5\frac{1}{4}$

   b. 3   e. NH

   c. $5\frac{3}{10}$

6. 9.201 x 7.0007 is closest to which number?

   f. 60
   g. 65
   h. 70
   j. 75
   k. NH

7. Which means the following?
   Seven more than three times a number (n).

   a. 7 + 3n
   b. 7 x 3n
   c. 3 + 7n
   d. 3n - 7
   e. NH

8. What is the area of this triangle?

   f. 20 m²
   g. 14 m²
   h. 23 m²
   j. 32 m²
   k. NH

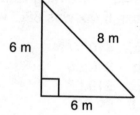

9. Which means the following?
   8 times the sum of a number and 6 is 56?

   a. 8n + 6 = 56
   b. 8 (n + 6) = 56
   c. 8n − 6 = 56
   d. 8n x 6 = 56
   e. NH

10. 6 meters =

    f. 6,000 kilometers
    g. 60 centimeters
    h. 600 millimeters
    j. 600 centimeters
    k. NH

Directions

Read each question and choose the correct answer. Mark the space for the answer you have chosen.
Mark NH if the answer is not here.

---

1. 700 millimeters =

   a. **7 kilometers**
   b. **7,000 meters**
   c. **7,000 centimeters**
   d. **70 centimeters**
   e. **NH**

10. If $\overline{AB}$ measures 18 inches, what is the measure of $\overline{DC}$?

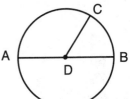

   f. **7 in**
   g. **8 in**
   h. **9 in**
   j. **18 in**
   k. **NH**

---

2. How would a 65° angle be classified?

   f. **acute**
   g. **right**
   h. **straight**
   j. **obtuse**
   k. **NH**

7. Sam bought seven 3¢ pieces of gum. How much change will he receive from a dollar? (There was no tax.)

   a. **1.00 + (7 x .03) = n**
   b. **1.00 ÷ (7 x .03) = n**
   c. **1.00 x (7 x .03) = n**
   d. **1.00 − (7 x .03) = n**
   e. **NH**

---

3. What is the area of a square with a side of 12 meters?

   a. **24 m²**
   b. **144 m²**
   c. **132 m²**
   d. **48 m²**
   e. **NH**

8. 6 kilometers =

   f. **6,000 meters**
   g. **600,000 meters**
   h. **60 meters**
   j. **60,000 meters**
   k. **NH**

---

4. Mark bought six magazines for $2.50 each. There was no tax. How much did he spend?

   f. **$12.00**
   g. **$15.00**
   h. **$12.50**
   j. **$8.60**
   k. **NH**

9. It is now 9:30 a.m. What time will it be in 2 hours and 10 minutes?

   a. **11:05 a.m.**
   b. **11:40 a.m.**
   c. **11:15 a.m.**
   d. **12:25 a.m.**
   e. **NH**

---

5. Steve spent 40% of his $500 in savings. How much money did Steve spend?

   a. **$200**
   b. **$260**
   c. **$220**
   d. **$280**
   e. **NH**

10. Which angle is a right angle?

   f.   h.

   g. ⟵————⟶   j.

            k. **NH**

Name _____

Directions

Read each question and choose the correct answer. Mark the space for the answer you have chosen. Mark NH if the answer is not here.

---

1. 9.125 x 6.887 is closest to which number?

    a. 50
    b. 60
    c. 70
    d. 80
    e. NH

6. What is the perimeter of this triangle?

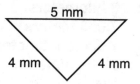

    f. 13 mm
    g. 90 mm
    h. 36 mm
    j. 25 mm
    k. NH

---

2. If Jack paid 90¢ for $\frac{1}{3}$ of a pound of oranges, what was the cost for a pound?

    f. $1.85
    g. $1.80
    h. $2.50
    j. $2.70
    k. NH

7. Which of the following is the measure of a right angle?

    a. 89°
    b. 87°
    c. 90°
    d. 91°
    e. NH

---

3. What is the perimeter of this square?

    a. 24 mm
    b. 12 mm   6 mm
    c. 36 mm
    d. 48 mm
    e. NH

8. A video game costs $35.95. If tax is $2.10, what is the total cost of the video game?

    f. $37.95
    g. $38.95
    h. $38.05
    j. $37.05
    k. NH

---

4. Sandy bought 5 ink pens for 19¢ each. How much change did he receive from a dollar? (There was no tax.)

    f. (19 x 1.00) = n
    g. 1.00 – (5 x .19) = n
    h. 1.00 + 5 + .19 = n
    j. 1.00 – (5 ÷ .19) = n
    k. NH

9. 900 grams =

    a. 9,000 kilograms
    b. 90 kilograms
    c. 9 kilograms
    d. 0.09 kilograms
    e. NH

---

5. What is the volume of a cube with a side of 4 feet?

    a. 36 ft³
    b. 16 ft³
    c. 256 ft³
    d. 144 ft³
    e. NH

10. The baseball game started at 6:00 p.m. and ended at 9:10 p.m. How long did the game last?

    f. 4 hours 10 minutes
    g. 3 hours 5 minutes
    h. 4 hours 20 minutes
    j. 3 hours 10 minutes
    k. NH

Directions

Read each question and choose the correct answer. Mark the space for the answer you have chosen. Mark NH if the answer is not here.

---

1. 6 x 421 =

   a. 3,314
   b. 2,216
   c. 3,426
   d. 2,526
   e. NH

2. 1,295 ÷ 7 =

   f. 165
   g. 142
   h. 185
   j. 178
   k. NH

3. 2,816 ÷ 8 is between which numbers?

   a. 100 and 200
   b. 200 and 300
   c. 300 and 400
   d. 400 and 500
   e. NH

4. 3.3 + 8.7 + 9.8 =

   f. 17.24
   g. 13.9
   h. 21.8
   j. 21.9
   k. NH

5. 68.32 − 41.705 =

   a. 25.351
   b. 25.354
   c. 26.615
   d. 26.359
   e. NH

6. Estimate the answer by rounding:
$$3.15 \times 6.7 =$$

   f. 20
   g. 23
   h. 21
   j. 18
   k. NH

7. Estimate the answer by rounding:
$$58.42 \div 9.2 =$$

   a. 8
   b. 5
   c. 6
   d. 7
   e. NH

8. Estimate the answer by rounding:
$$253.3 \div 4.3 =$$

   f. 60
   g. 50
   h. 70
   j. 80
   k. NH

9. $\frac{3}{5} + \frac{2}{3} =$

   a. $1\frac{4}{12}$ d. $1\frac{2}{5}$

   b. $1\frac{5}{12}$ e. NH

   c. $1\frac{5}{8}$

10. $\frac{6}{5} - \frac{3}{7} =$

   f. $\frac{3}{2}$ j. $\frac{2}{21}$

   g. $\frac{3}{35}$ k. NH

   h. $\frac{27}{35}$

Name _____

Directions

Read each question and choose the correct answer. Mark the space for the answer you have chosen. Mark NH if the answer is not here.

---

1. 8 x 456 =

   a. 4,488
   b. 4,648
   c. 3,488
   d. 3,648
   e. NH

6. Estimate the answer by rounding:
        8.35 x 6.2 =

   f. 48
   g. 58
   h. 52
   j. 42
   k. NH

---

2. 1,266 ÷ 6 =

   f. 211
   g. 209
   h. 210
   j. 208
   k. NH

7. Estimate the answer by rounding:
        52.21 ÷ 8.7 =

   a. 8
   b. 6
   c. 5
   d. 7
   e. NH

---

3. 5,685 ÷ 12 is between which numbers?

   a. 100 and 200
   b. 200 and 300
   c. 300 and 400
   d. 400 and 500
   e. NH

8. Estimate the answer by rounding:
        221.3 ÷ 6.8 =

   f. 60
   g. 50
   h. 40
   j. 30
   k. NH

---

4. 4.8 + 6.3 + 8.5 =

   f. 19.2
   g. 19.5
   h. 19.6
   j. 19.9
   k. NH

9. $\frac{1}{3} + \frac{4}{6} =$

   a. 1   d. $3\frac{3}{5}$

   b. $1\frac{6}{12}$ e. NH

   c. $2\frac{1}{4}$

---

5. 38.15 − 21.645 =

   a. 16.351
   b. 16.505
   c. 16.349
   d. 16.359
   e. NH

10. $\frac{7}{8} - \frac{3}{8} =$

   f. $\frac{1}{6}$ j. $\frac{4}{21}$

   g. $\frac{5}{18}$ k. NH

   h. $\frac{5}{21}$

Name _____

Directions
Use the graph to answer the questions. Read each question and choose the correct answer. Mark the space for the answer you have chosen. Mark NH if the answer is not here.

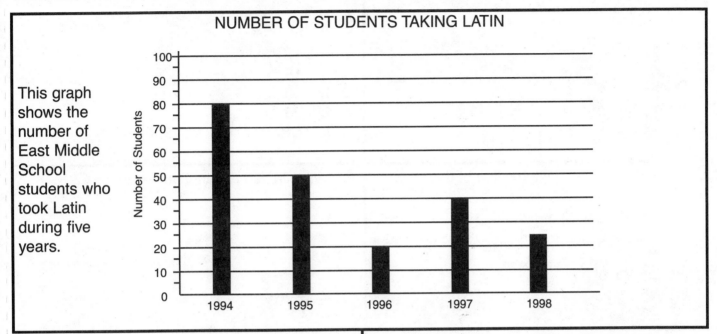

NUMBER OF STUDENTS TAKING LATIN

This graph shows the number of East Middle School students who took Latin during five years.

1. In what year did the most students take Latin?

   a. 1994
   b. 1995
   c. 1996
   d. 1998
   e. NH

2. About how many students took Latin in the years 1995 and 1997 together?

   f. 130
   g. 120
   h. 90
   j. 75
   k. NH

3. About how many students took Latin in the five years shown on the graph?

   a. 225
   b. 215
   c. 210
   d. 200
   e. NH

4. How many more students took Latin in 1994 than in 1996?

   f. 30
   g. 40
   h. 50
   j. 60
   k. NH

5. What two years had the fewest students taking Latin?

   a. 1994 and 1995
   b. 1995 and 1997
   c. 1996 and 1998
   d. 1997 and 1998
   e. NH

6. What was the average number of students taking Latin during the five years shown?

   f. 43
   g. 72
   h. 45
   j. 36
   k. NH

Directions

Read each question and choose the correct answer. Mark the space for the answer you have chosen. Mark NH if the answer is not here.

1. Which is a prime number?

   a. 4
   b. 7
   c. 8
   d. 9
   e. NH

2. What is the prime factorization of 30?

   f. 5 x 6
   g. 2 x 2 x 5
   h. 15 x 2
   j. 2 x 3 x 5
   k. NH

3. Mabel has saved $20 toward a $40 book. What fraction of the price has she saved?

   a. $\frac{2}{5}$ d. $\frac{1}{3}$

   b. $\frac{2}{3}$ e. NH

   c. $\frac{1}{2}$

4. 810 ÷ 20 is between which numbers?

   f. 50 and 60
   g. 40 and 50
   h. 30 and 40
   j. 20 and 30
   k. NH

5. 20 is 50% of what number?

   a. 30
   b. 40
   c. 60
   d. 100
   e. NH

6. If $\frac{4}{10} = \frac{6}{x}$, then x =

   f. 12
   g. 15
   h. 18
   j. 20
   k. NH

7. What fraction is another name for $\frac{2}{7}$?

   a. $\frac{4}{21}$ d. $\frac{4}{14}$

   b. $\frac{8}{24}$ e. NH

   c. $\frac{3}{20}$

8. 8.05 − 4.99 =

   f. 0.306
   g. 3.06
   h. 30.6
   j. 306
   k. NH

9. What is the numeral for ten million, one hundred forty-one?

   a. 10,141
   b. 10,100,041
   c. 10,141,000
   d. 10,000,141
   e. NH

10. 53 x 107 =

   f. 5,671
   g. 521
   h. 5,321
   j. 5,171
   k. NH

Directions

Read each question and choose the correct answer. Mark the space for the answer you have chosen. Mark NH if the answer is not here.

---

1. $\frac{2}{3} \times \frac{6}{8} =$

   a. $\frac{1}{3}$ **d.** $\frac{2}{3}$

   b. $\frac{1}{6}$ **e.** NH

   c. $\frac{1}{2}$

6. 8 is 10% of what number?

   f. 880
   g. 80
   h. 88
   j. 800
   k. NH

---

2. What is the lowest common denominator for $\frac{3}{5}$ and $\frac{4}{7}$?

   f. 28
   g. 35
   h. 42
   j. 60
   k. NH

7. Solve for n:
$$n - 7 = 11?$$

   a. 4
   b. 7
   c. 16
   d. 18
   e. NH

---

3. Which is a prime number?

   a. 72
   b. 51
   c. 18
   d. 17
   e. NH

8. If $n + 6 = 12$, then $n =$

   f. 7
   g. 6
   h. 5
   j. 4
   k. NH

---

4. Which number has the smallest value?

   f. 0.453
   g. 0.345
   h. 0.354
   j. 0.435
   k. NH

9. What is 425.6 rounded to the nearest whole number?

   a. 430
   b. 425
   c. 426
   d. 424
   e. NH

---

5. What number is 7% of 300?

   a. 210
   b. 2.1
   c. 21
   d. 0.21
   e. NH

10. What number is 2 hundredths more than 5.063?

   f. 5.083
   g. 5.065
   h. 7.063
   j. 5.263
   k. NH

Name _____

Directions

Use the graph to answer the questions. Read each question and choose the correct answer. Mark the space for the answer you have chosen. Mark NH if the answer is not here.

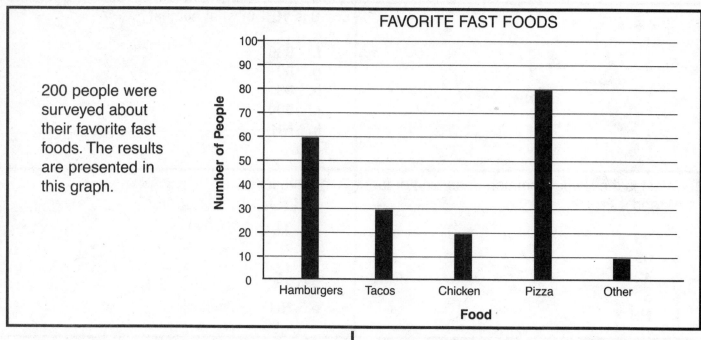

200 people were surveyed about their favorite fast foods. The results are presented in this graph.

FAVORITE FAST FOODS

Number of People

Hamburgers   Tacos   Chicken   Pizza   Other

Food

1. How many of the people chose hamburgers as their favorite fast food?

   a. 60
   b. 70
   c. 80
   d. 90
   e. NH

4. How many people chose either hamburgers or tacos as their favorite fast food?

   f. 60
   g. 70
   h. 80
   j. 90
   k. NH

2. How many people chose tacos or chicken as their favorite fast food?

   f. 40
   g. 50
   h. 60
   j. 70
   k. NH

5. How many people did not choose pizza as their favorite fast food?

   a. 80
   b. 120
   c. 140
   d. 160
   e. NH

3. How many more people chose tacos than chicken as their favorite fast food?

   a. 40
   b. 30
   c. 20
   d. 10
   e. NH

6. How many more people chose pizza than tacos as their favorite food?

   f. 20
   g. 30
   h. 40
   j. 50
   k. NH

Directions

Read each question and choose the correct answer. Mark the space for the answer you have chosen. Mark NH if the answer is not here.

---

1. $\frac{1}{2} \div 2 =$

   a. 1   d. $\frac{1}{2}$

   b. 2   e. NH

   c. $\frac{1}{4}$

6. What is the prime factorization of 90?

   f. 9 x 2 x 5
   g. 2 x 3 x 3 x 5
   h. 2 x 45
   j. 6 x 3 x 5
   k. NH

---

2. What does 9 in 0.19 represent?

   f. 9 thousandths
   g. 9 ten thousandths
   h. 9 hundredths
   j. 9 tenths
   k. NH

7. What is 35% written as a fraction in its lowest terms?

   a. $\frac{35}{100}$ d. $\frac{8}{25}$

   b. $\frac{7}{20}$ e. NH

   c. $\frac{6}{10}$

---

3. What is 5,625 rounded to the nearest hundred?

   a. 5,700
   b. 6,000
   c. 5,600
   d. 5,630
   e. NH

8. Sue bought $\frac{1}{3}$ of a pound of nuts for 90¢. What is the cost per pound for nuts?

   f. 30¢
   g. $1.50
   h. $2.10
   j. $2.70
   k. NH

---

4. What is the least common denominator for $\frac{2}{3}$ and $\frac{4}{5}$?

   f. 6
   g. 9
   h. 10
   j. 15
   k. NH

9. We are in school from 8:30 a.m. until 3:30 p.m. How long are we in school?

   a. 6 hours
   b. 6 hours and 30 minutes
   c. 7 hours
   d. 7 hours and 30 minutes
   e. NH

---

5. Which is a prime number?

   a. 29
   b. 4
   c. 14
   d. 27
   e. NH

10. Joe received 20% of the 1,400 votes that were cast. How many votes did Joe receive?

   f. 240
   g. 260
   h. 280
   j. 400
   k. NH

Directions

Read each question and choose the correct answer. Mark the space for the answer you have chosen. Mark NH if the answer is not here.

---

1. 800 grams =

   a. **8,000 kilograms**
   b. **80 kilograms**
   c. **8 kilograms**
   d. **0.8 kilograms**
   e. **NH**

2. Bess had $20.44 in her bank. She took out $4.95 to buy a new book. How much money is left in her bank?
   f. **$15.49**
   g. **$16.51**
   h. **$16.59**
   j. **$15.59**
   k. **NH**

3. Jim charges $4 per hour for baby-sitting. He sat 2 hours on Friday and 6 hours on Saturday. How much money did he make?
   a. **$8.00**
   b. **$12.00**
   c. **$24.00**
   d. **$32.00**
   e. **NH**

4. Sam had 27 hits in the last 9 games. How many hits does he average per game?

   f. **16**
   g. **9**
   h. **4**
   j. **3**
   k. **NH**

5. If it takes 4 hours to cook a 6-pound roast, how long will it take to cook an 18-pound roast?

   a. **10 hours**
   b. **11 hours**
   c. **12 hours**
   d. **16 hours**
   e. **NH**

6. 15 is what percent of 60?

   f. **20%**
   g. **25%**
   h. **30%**
   j. **35%**
   k. **NH**

7. Which number sentence means, "The quotient of a number (n) and six"?

   a. **n ÷ 6**
   b. **6n**
   c. **n + 6**
   d. **n − 6**
   e. **NH**

8. Which of the following could <u>not</u> be the measure of an acute angle?

   f. **42°**
   g. **59°**
   h. **99°**
   j. **25°**
   k. **NH**

9. What is the measure of angle RTS?

   a. **90°**
   b. **80°**
   c. **70°**
   d. **60°**
   e. **NH**

10. What is the perimeter of this square?

   f. **18 m**
   g. **27 m**
   h. **39 m**
   j. **45 m**
   k. **NH**

Answer Key

	Page 1	Page 2	Page 3	Page 4	Page 5	Page 6
1	d	b	a	b	b	a
2	f	f	h	g	k	g
3	b	d	b	a	b	e
4	h	g	g	j	f	g
5	d	c	e	c	c	e
6	g	h	f	j	g	h
7	c	b	e	a	c	b
8	j	f	g	f	j	h
9	a	d	b	d	d	d
10	g	f	g	h	g	g

	Page 7	Page 8	Page 9	Page 10	Page 11	Page 12
1	b	c	b	b	d	b
2	f	f	g	f	g	j
3	c	b	a	b	a	c
4	j	j	g	j	g	j
5	b	d	d	c	c	b
6	g	h	g	h	f	g
7	d	a	a	c	c	b
8	g	j	j	j	h	f
9	d	b	b	d	d	c
10	g	g	g	h	h	f

	Page 13	Page 14	Page 15	Page 16	Page 17	Page 18
1	c	b	a	d	c	d
2	j	f	h	f	g	f
3	d	b	b	b	b	b
4	g	g	g	j	j	g
5	a	c	a	c	c	c
6	f	f	h	g	g	g
7	b	b	c	b	d	c
8	f	j	h	j	g	k
9	c	d	d	a	a	b
10	h	f	g	f	f	f

Answer Key

	Page 19	Page 20	Page 21	Page 22	Page 23	Page 24
1	c	b	c	c	b	a
2	h	g	g	f	f	h
3	a	a	a	b	d	d
4	g	j	h	j	g	k
5	d	c	c	b	c	b
6	g	j	j	j	h	j
7	c	d	b	a	b	b
8	f	f	f	f	j	k
9	c	c	d	c	b	c
10	j	g	g	g	g	h

	Page 25	Page 26	Page 27	Page 28	Page 29	Page 30
1	d	a	b	c	a	a
2	h	g	h	f	h	h
3	a	c	a	b	c	d
4	j	f	g	j	f	f
5	b	b	a	a	a	b
6	h	f	g	g	h	j
7	e	b	d	c	c	c
8	g	g	k	j	h	g
9	d	b	c	b	d	b
10	h	h	g	g	g	j

	Page 31	Page 32	Page 33	Page 34	Page 35	Page 36
1	a	a	a	c	b	a
2	h	h	g	h	f	h
3	b	b	b	a	b	c
4	f	j	f	h	f	j
5	c	a	c	d	c	a
6	j	j	j	h	f	g
7	a	b	b	c	b	b
8	h	f	j	j	j	f
9	b	c	d	b	b	b
10	f	f	g	k	f	j

Answer Key

Page 37
1. ● d
2. ● g
3. ● c
4. ● g
5. ● b
6. ● h
7. ● c
8. ● j
9. ● b
10. ● f

Page 38
1. ● e
2. ● j
3. ● b
4. ● h
5. ● b
6. ● k
7. ● c
8. ● h
9. ● c
10. ● f

Page 39
1. ● b
2. ● h
3. ● a
4. ● h
5. ● c
6. ● j
7. ● b
8. ● j
9. ● a
10. ● g

Page 40
1. ● b
2. ● g
3. ● b
4. ● f
5. ● d
6. ● g
7. ● b
8. ● j
9. ● c
10. ● f

Page 41
1. ● b
2. ● j
3. ● b
4. ● g
5. ● c
6. ● g
7. ● a
8. ● f
9. ● d
10. ● g

Page 42
1. ● c
2. ● f
3. ● d
4. ● f
5. ● b
6. ● h
7. ● a
8. ● h
9. ● b
10. ● g

Page 43
1. ● e
2. ● g
3. ● a
4. ● g
5. ● d
6. ● f
7. ● a
8. ● h
9. ● c
10. ● f

Page 44
1. ● c
2. ● f
3. ● b
4. ● k
5. ● d
6. ● g
7. ● c
8. ● f
9. ● k
10. ● g

Page 45
1. ● d
2. ● g
3. ● a
4. ● j
5. ● d
6. ● g
7. ● b
8. ● h
9. ● f
10. ● g

Page 46
1. ● c
2. ● h
3. ● e
4. ● h
5. ● d
6. ● g
7. ● d
8. ● h
9. ● c
10. ● j

Page 47
1. ● d
2. ● g
3. ● c
4. ● f
5. ● a
6. ● f
7. ● b
8. ● g
9. ● c
10. ● g

Page 48
1. ● c
2. ● f
3. ● a
4. ● j
5. ● a
6. ● g
7. ● c
8. ● f
9. ● b
10. ● h

Page 49
1. ● a
2. ● h
3. ● b
4. ● j
5. ● a
6. ● h
7. ● c
8. ● f
9. ● b
10. ● g

Page 50
1. ● e
2. ● f
3. ● b
4. ● h
5. ● a
6. ● h
7. ● e
8. ● f
9. ● b
10. ● g

Page 51
1. ● b
2. ● k
3. ● b
4. ● g
5. ● c
6. ● h
7. ● d
8. ● g
9. ● c
10. ● g

Page 52
1. ● a
2. ● g
3. ● b
4. ● k
5. ● b
6. ● h
7. ● d
8. ● g
9. ● c
10. ● g

Page 53
1. ● b
2. ● k
3. ● d
4. ● h
5. ● b
6. ● h
7. ● e
8. ● g
9. ● e
10. ● g

Page 54
1. ● b
2. ● g
3. ● d
4. ● h
5. ● d
6. ● j
7. ● d
8. ● g
9. ● b
10. ● g

Answer Key

Page 55	Page 56	Page 57	Page 58	Page 59	Page 60
1. a	1. b	1. d	1. d	1. b	1. b
2. h	2. h	2. g	2. f	2. h	2. h
3. a	3. d	3. a	3. c	3. a	3. b
4. g	4. j	4. h	4. g	4. j	4. h
5. a	5. a	5. d	5. b	5. c	5. d
6. j	6. j	6. j	6. f	6. h	6. j
7. b	7. c	7. a	7. a	7. c	7. a
8. g	8. f	8. h	8. h	8. f	8. j
9. a	9. c	9. b	9. c	9. b	9. a
10. h	10. h	10. f	10. f	10. f	10. h

Page 61	Page 62	Page 63	Page 64	Page 65	Page 66
1. c	1. c	1. c	1. b	1. a	1. b
2. f	2. h	2. h	2. j	2. j	2. j
3. d	3. a	3. b	3. b	3. b	3. b
4. f	4. j	4. h	4. k	4. h	4. h
5. c	5. b	5. c	5. a	5. b	5. a
6. f	6. h	6. f	6. j	6. g	6. h
7. e	7. d	7. c	7. b	7. b	7. c
8. h	8. g	8. g	8. g	8. g	8. f
9. b	9. c	9. a	9. d	9. c	9. c
10. f	10. f	10. f	10. g	10. f	10. g

Page 67	Page 68	Page 69	Page 70	Page 71	Page 72
1. b	1. a	1. b	1. b	1. c	1. b
2. g	2. g	2. g	2. g	2. f	2. g
3. c	3. b	3. c	3. d	3. c	3. a
4. g	4. h	4. h	4. f	4. g	4. f
5. a	5. d	5. c	5. e	5. a	5. b
6. j	6. j	6. h	6. h	6. h	6. h
7. b	7. c	7. b	7. b	7. b	7. c
8. g	8. h	8. j	8. j	8. f	8. j
9. c	9. c	9. d	9. a	9. b	9. d
10. j	10. g	10. g	10. j	10. f	10. g

Answer Key

Page 73
1. b
2. g
3. a
4. j
5. b
6. j
7. a
8. j
9. d
10. g

Page 74
1. e
2. f
3. e
4. g
5. c
6. j
7. a
8. h

Page 75
1. a
2. h
3. b
4. f
5. d
6. j

Page 76
1. c
2. h
3. c
4. h

Page 77
1. d
2. f
3. b
4. j
5. d
6. g
7. d
8. g

Page 78
1. a
2. g
3. b
4. j
5. c
6. g
7. c
8. j

Page 79
1. c
2. h
3. a
4. j
5. d
6. g
7. b
8. h

Page 80
1. a
2. h
3. c
4. f
5. d
6. g
7. d
8. h

Page 81
1. a
2. h
3. a
4. h
5. d
6. f
7. d
8. f

Page 82
1. b
2. j
3. a
4. j
5. a
6. h
7. d
8. h

Page 83
1. d
2. g
3. b
4. f
5. c
6. j
7. b
8. j

Page 84
1. c
2. g
3. b
4. f
5. c
6. h
7. d
8. h

Page 85
1. d
2. j
3. c
4. h
5. d
6. g
7. a
8. f

Page 86
1. b
2. j
3. d
4. f
5. a
6. h
7. d
8. h

Page 87
1. b
2. f
3. d
4. g
5. b
6. j
7. b
8. h

Page 88
1. b
2. h
3. c
4. g
5. d
6. g
7. d
8. g

Page 89
1. a
2. j
3. d
4. h
5. b
6. j
7. b
8. h

Page 90
1. a
2. g
3. a
4. g
5. a
6. f
7. b
8. g

Page 91
1. d
2. f
3. b
4. g
5. d
6. k
7. b

Page 92
1. a
2. h
3. c
4. h
5. d
6. f
7. a

Page 93
1. c
2. h
3. b
4. f
5. b
6. g
7. a
8. g

Page 94
1. a
2. h
3. c
4. f
5. a
6. g
7. c
8. f

Page 95
1. a
2. h
3. b
4. j
5. b
6. f
7. a
8. j

Page 96
1. b
2. h
3. a
4. f
5. c
6. j
7. d
8. h

Answer Key

Page 97
1. e
2. j
3. b
4. g
5. a
6. k
7. b
8. j

Page 98
1. a
2. j
3. a
4. k
5. a
6. k

Page 99
1. d
2. g
3. c
4. j
5. a
6. g

Page 100
1. c
2. j
3. b
4. k
5. a
6. k

Page 101
1. c
2. j
3. b
4. j
5. a
6. f
7. e
8. g
9. a
10. j

Page 102
1. c
2. f
3. a
4. j
5. a
6. h
7. c
8. k
9. b
10. j

Page 103
1. d
2. f
3. c
4. h
5. c
6. g
7. d
8. f
9. b
10. g

Page 104
1. d
2. g
3. c
4. g
5. c
6. g
7. d
8. j
9. c
10. j

Page 105
1. b
2. g
3. c
4. j
5. b
6. j
7. b
8. h
9. a
10. g

Page 106
1. b
2. g
3. a
4. j
5. e
6. h
7. a
8. j
9. a
10. k

Page 107
1. d
2. g
3. b
4. h
5. a
6. g
7. d
8. g
9. c
10. j

Page 108
1. b
2. j
3. a
4. h
5. e
6. g
7. c
8. h
9. e
10. j

Page 109
1. d
2. j
3. c
4. h
5. c
6. h
7. d
8. f
9. e
10. h

Page 110
1. d
2. f
3. d
4. h
5. b
6. k
7. b
8. j
9. f
10. k

Page 111
1. a
2. h
3. b
4. k
5. c
6. f

Page 112
1. b
2. j
3. c
4. h
5. b
6. h
7. d
8. h
9. d
10. f

Page 113
1. c
2. f
3. d
4. g
5. b
6. g
7. d
8. g
9. c
10. f

Page 114
1. a
2. h
3. d
4. j
5. c
6. j

Page 115
1. c
2. h
3. c
4. j
5. a
6. g
7. b
8. j
9. c
10. h

Page 116
1. d
2. f
3. d
4. j
5. c
6. g
7. a
8. h
9. a
10. k

Answer Sheet

Name _____

Page _____
Score _____
1. ⓐ ⓑ ⓒ ⓓ ⓔ
2. ⓕ ⓖ ⓗ ⓙ ⓚ
3. ⓐ ⓑ ⓒ ⓓ ⓔ
4. ⓕ ⓖ ⓗ ⓙ ⓚ
5. ⓐ ⓑ ⓒ ⓓ ⓔ
6. ⓕ ⓖ ⓗ ⓙ ⓚ
7. ⓐ ⓑ ⓒ ⓓ ⓔ
8. ⓕ ⓖ ⓗ ⓙ ⓚ
9. ⓐ ⓑ ⓒ ⓓ ⓔ
10. ⓕ ⓖ ⓗ ⓙ ⓚ
11. ⓐ ⓑ ⓒ ⓓ ⓔ
12. ⓕ ⓖ ⓗ ⓙ ⓚ

Page _____
Score _____
1. ⓐ ⓑ ⓒ ⓓ ⓔ
2. ⓕ ⓖ ⓗ ⓙ ⓚ
3. ⓐ ⓑ ⓒ ⓓ ⓔ
4. ⓕ ⓖ ⓗ ⓙ ⓚ
5. ⓐ ⓑ ⓒ ⓓ ⓔ
6. ⓕ ⓖ ⓗ ⓙ ⓚ
7. ⓐ ⓑ ⓒ ⓓ ⓔ
8. ⓕ ⓖ ⓗ ⓙ ⓚ
9. ⓐ ⓑ ⓒ ⓓ ⓔ
10. ⓕ ⓖ ⓗ ⓙ ⓚ
11. ⓐ ⓑ ⓒ ⓓ ⓔ
12. ⓕ ⓖ ⓗ ⓙ ⓚ

Page _____
Score _____
1. ⓐ ⓑ ⓒ ⓓ ⓔ
2. ⓕ ⓖ ⓗ ⓙ ⓚ
3. ⓐ ⓑ ⓒ ⓓ ⓔ
4. ⓕ ⓖ ⓗ ⓙ ⓚ
5. ⓐ ⓑ ⓒ ⓓ ⓔ
6. ⓕ ⓖ ⓗ ⓙ ⓚ
7. ⓐ ⓑ ⓒ ⓓ ⓔ
8. ⓕ ⓖ ⓗ ⓙ ⓚ
9. ⓐ ⓑ ⓒ ⓓ ⓔ
10. ⓕ ⓖ ⓗ ⓙ ⓚ
11. ⓐ ⓑ ⓒ ⓓ ⓔ
12. ⓕ ⓖ ⓗ ⓙ ⓚ

Page _____
Score _____
1. ⓐ ⓑ ⓒ ⓓ ⓔ
2. ⓕ ⓖ ⓗ ⓙ ⓚ
3. ⓐ ⓑ ⓒ ⓓ ⓔ
4. ⓕ ⓖ ⓗ ⓙ ⓚ
5. ⓐ ⓑ ⓒ ⓓ ⓔ
6. ⓕ ⓖ ⓗ ⓙ ⓚ
7. ⓐ ⓑ ⓒ ⓓ ⓔ
8. ⓕ ⓖ ⓗ ⓙ ⓚ
9. ⓐ ⓑ ⓒ ⓓ ⓔ
10. ⓕ ⓖ ⓗ ⓙ ⓚ
11. ⓐ ⓑ ⓒ ⓓ ⓔ
12. ⓕ ⓖ ⓗ ⓙ ⓚ

Page _____
Score _____
1. ⓐ ⓑ ⓒ ⓓ ⓔ
2. ⓕ ⓖ ⓗ ⓙ ⓚ
3. ⓐ ⓑ ⓒ ⓓ ⓔ
4. ⓕ ⓖ ⓗ ⓙ ⓚ
5. ⓐ ⓑ ⓒ ⓓ ⓔ
6. ⓕ ⓖ ⓗ ⓙ ⓚ
7. ⓐ ⓑ ⓒ ⓓ ⓔ
8. ⓕ ⓖ ⓗ ⓙ ⓚ
9. ⓐ ⓑ ⓒ ⓓ ⓔ
10. ⓕ ⓖ ⓗ ⓙ ⓚ
11. ⓐ ⓑ ⓒ ⓓ ⓔ
12. ⓕ ⓖ ⓗ ⓙ ⓚ

Page _____
Score _____
1. ⓐ ⓑ ⓒ ⓓ ⓔ
2. ⓕ ⓖ ⓗ ⓙ ⓚ
3. ⓐ ⓑ ⓒ ⓓ ⓔ
4. ⓕ ⓖ ⓗ ⓙ ⓚ
5. ⓐ ⓑ ⓒ ⓓ ⓔ
6. ⓕ ⓖ ⓗ ⓙ ⓚ
7. ⓐ ⓑ ⓒ ⓓ ⓔ
8. ⓕ ⓖ ⓗ ⓙ ⓚ
9. ⓐ ⓑ ⓒ ⓓ ⓔ
10. ⓕ ⓖ ⓗ ⓙ ⓚ
11. ⓐ ⓑ ⓒ ⓓ ⓔ
12. ⓕ ⓖ ⓗ ⓙ ⓚ

Answer Sheet

Name _____

Page _____
Score _____

1. (a) (b) (c) (d) (e)
2. (f) (g) (h) (j) (k)
3. (a) (b) (c) (d) (e)
4. (f) (g) (h) (j) (k)
5. (a) (b) (c) (d) (e)
6. (f) (g) (h) (j) (k)
7. (a) (b) (c) (d) (e)
8. (f) (g) (h) (j) (k)
9. (a) (b) (c) (d) (e)
10. (f) (g) (h) (j) (k)
11. (a) (b) (c) (d) (e)
12. (f) (g) (h) (j) (k)

Page _____
Score _____

1. (a) (b) (c) (d) (e)
2. (f) (g) (h) (j) (k)
3. (a) (b) (c) (d) (e)
4. (f) (g) (h) (j) (k)
5. (a) (b) (c) (d) (e)
6. (f) (g) (h) (j) (k)
7. (a) (b) (c) (d) (e)
8. (f) (g) (h) (j) (k)
9. (a) (b) (c) (d) (e)
10. (f) (g) (h) (j) (k)
11. (a) (b) (c) (d) (e)
12. (f) (g) (h) (j) (k)

Page _____
Score _____

1. (a) (b) (c) (d) (e)
2. (f) (g) (h) (j) (k)
3. (a) (b) (c) (d) (e)
4. (f) (g) (h) (j) (k)
5. (a) (b) (c) (d) (e)
6. (f) (g) (h) (j) (k)
7. (a) (b) (c) (d) (e)
8. (f) (g) (h) (j) (k)
9. (a) (b) (c) (d) (e)
10. (f) (g) (h) (j) (k)
11. (a) (b) (c) (d) (e)
12. (f) (g) (h) (j) (k)

Page _____
Score _____

1. (a) (b) (c) (d) (e)
2. (f) (g) (h) (j) (k)
3. (a) (b) (c) (d) (e)
4. (f) (g) (h) (j) (k)
5. (a) (b) (c) (d) (e)
6. (f) (g) (h) (j) (k)
7. (a) (b) (c) (d) (e)
8. (f) (g) (h) (j) (k)
9. (a) (b) (c) (d) (e)
10. (f) (g) (h) (j) (k)
11. (a) (b) (c) (d) (e)
12. (f) (g) (h) (j) (k)

Page _____
Score _____

1. (a) (b) (c) (d) (e)
2. (f) (g) (h) (j) (k)
3. (a) (b) (c) (d) (e)
4. (f) (g) (h) (j) (k)
5. (a) (b) (c) (d) (e)
6. (f) (g) (h) (j) (k)
7. (a) (b) (c) (d) (e)
8. (f) (g) (h) (j) (k)
9. (a) (b) (c) (d) (e)
10. (f) (g) (h) (j) (k)
11. (a) (b) (c) (d) (e)
12. (f) (g) (h) (j) (k)

Page _____
Score _____

1. (a) (b) (c) (d) (e)
2. (f) (g) (h) (j) (k)
3. (a) (b) (c) (d) (e)
4. (f) (g) (h) (j) (k)
5. (a) (b) (c) (d) (e)
6. (f) (g) (h) (j) (k)
7. (a) (b) (c) (d) (e)
8. (f) (g) (h) (j) (k)
9. (a) (b) (c) (d) (e)
10. (f) (g) (h) (j) (k)
11. (a) (b) (c) (d) (e)
12. (f) (g) (h) (j) (k)

4	17.6
right	acute
3	54
22 m	6 sq ft
13.5	9
obtuse	7
6	5
28 sq cm	14 mi

Find the percentage.

5% of 80 =

Find the percentage.

16% of 110 =

Find the percentage.

45% of 30 =

Find the percentage.

15% of 60 =

Identify this angle as right, obtuse, or acute.

Identify this angle as right, obtuse, or acute.

Identify this angle as right, obtuse, or acute.

If $n + 2 = 9$, then n =

If $n + 9 = 12$, then n =

If $n = 9$, then $n \times 6 =$

If $n = 12$, then $n \div 2 =$

If $n = 20$, then $n - 15 =$

Find the perimeter.

4 m · 7 m · 7 m · 4 m

Find the area.

2 ft · 3 ft

Find the area.

4 cm · 7 cm

Find the perimeter.

1 mi · 6 mi · 6 mi · 1 mi

17.8	5.2	$\frac{4}{5}$	$\frac{17}{25}$
$6\frac{1}{2}$	$75\frac{1}{5}$	0.625	0.55
$\frac{4}{5}$	$\frac{3}{20}$	$\frac{17}{20}$	$\frac{1}{5}$
40%	35%	20%	20%

Divide.

$35.6 \div 2 =$

Divide.

$82.16 \div 15.8 =$

Change to a fraction and reduce to lowest terms.

6.5

Change to a fraction and reduce to lowest terms.

75.2

Change to a fraction and reduce to lowest terms.

0.8

Change to a decimal.

$\dfrac{5}{8}$

Change to a decimal.

0.68

Change to a fraction and reduce to lowest terms.

$\dfrac{11}{20}$

80%

Change to a percent.

15%

Change to a percent.

85%

Change to a percent.

20%

Change to a percent.

Change to a percent.

$\dfrac{2}{5}$

Change to a percent.

$\dfrac{7}{20}$

Change to a percent.

$\dfrac{5}{25}$

Change to a percent.

$\dfrac{10}{50}$

$\dfrac{1}{3}$

29.97

23.78

21.775

© CD-3756

$2\dfrac{5}{12}$

123.92

35.319

2,473.8

© CD-3756

30.7

72.2

16

0.6

© CD-3756

2.303

294.95

2.9748

0.173

© CD-3756

Subtract and reduce to simplest terms.

$$2\frac{1}{6} - 1\frac{5}{6} =$$

Subtract and reduce to simplest terms.

$$5\frac{2}{3} - 3\frac{1}{4} =$$

Add.

$$14.2 + 16.5 =$$

Add.

$$0.603 + 1.7 =$$

Subtract.

$$13.87 + 16.1 =$$

Add.

$$0.52 + 123.4 =$$

Subtract.

$$75.4 - 3.2 =$$

Subtract.

$$306.45 - 11.5 =$$

Subtract.

$$28.4 - 4.62 =$$

Subtract.

$$42.619 - 7.3 =$$

Multiply.

$$6.4 \times 2.5 =$$

Multiply.

$$4.44 \times .67 =$$

Multiply.

$$87.1 \times 0.25 =$$

Multiply.

$$9.5 \times 260.4 =$$

Divide.

$$3.6 \div 6 =$$

Divide.

$$0.865 \div 5 =$$

$\frac{1}{2}$ © CD-3756	$1\frac{1}{3}$ © CD-3756	14 © CD-3756	$22\frac{11}{12}$ © CD-3756
$\frac{3}{10}$ © CD-3756	$4\frac{1}{6}$ © CD-3756	$\frac{1}{2}$ © CD-3756	1 © CD-3756
14 © CD-3756	$\frac{3}{5}$ © CD-3756	$\frac{8}{9}$ © CD-3756	$13\frac{19}{30}$ © CD-3756
$\frac{4}{7}$ © CD-3756	$\frac{3}{7}$ © CD-3756	$3\frac{1}{3}$ © CD-3756	$6\frac{3}{5}$ © CD-3756

Multiply and reduce to simplest terms.

$$\frac{3}{4} \times \frac{2}{3} =$$

Multiply and reduce to simplest terms.

$$\frac{3}{5} \times \frac{1}{2} =$$

Add and reduce to simplest terms.

$$6\frac{2}{5} + 7\frac{3}{5} =$$

Subtract and reduce to simplest terms.

$$\frac{9}{14} - \frac{1}{14} =$$

Multiply and reduce to simplest terms.

$$\frac{3}{9} \times 4 =$$

Multiply and reduce to simplest terms.

$$\frac{5}{6} \times 5 =$$

Add and reduce to simplest terms.

$$\frac{1}{10} + \frac{4}{8} =$$

Subtract and reduce to simplest terms.

$$\frac{5}{7} - \frac{2}{7} =$$

Multiply and reduce to simplest terms.

$$3\frac{1}{2} \times 4 =$$

Add and reduce to simplest terms.

$$\frac{3}{8} + \frac{1}{8} =$$

Add and reduce to simplest terms.

$$\frac{2}{9} + \frac{2}{3} =$$

Subtract and reduce to simplest terms.

$$4 - \frac{2}{3} =$$

Multiply and reduce to simplest terms.

$$6\frac{7}{8} \times 3\frac{1}{3} =$$

Add and reduce to simplest terms.

$$\frac{1}{2} + \frac{1}{2} =$$

Add and reduce to simplest terms.

$$6\frac{3}{10} + 7\frac{1}{3} =$$

Subtract and reduce to simplest terms.

$$8\frac{4}{5} - 2\frac{1}{5} =$$

$\dfrac{3}{4}$ 	$\dfrac{1}{5}$ 	$\dfrac{1}{2}$ 	$\dfrac{3}{4}$
$\dfrac{21}{4}$ 	$\dfrac{19}{8}$ 	$\dfrac{31}{5}$ 	$\dfrac{26}{3}$
$1\dfrac{1}{3}$ 	$2\dfrac{2}{3}$ 	$3\dfrac{1}{4}$ 	$6\dfrac{1}{7}$
$\dfrac{2}{3}=\dfrac{8}{12}$ 	$\dfrac{3}{7}=\dfrac{6}{14}$ 	$\dfrac{2}{3}=\dfrac{6}{9}$ 	$\dfrac{1}{6}=\dfrac{6}{36}$

Change to simplest form.

$$\frac{6}{8}$$

Change to simplest form.

$$\frac{5}{25}$$

Change to simplest form.

$$\frac{15}{30}$$

Change to simplest form.

$$\frac{24}{32}$$

Change to a fraction.

$$5\frac{1}{4}$$

Change to a fraction.

$$2\frac{3}{8}$$

Change to a fraction.

$$6\frac{1}{5}$$

Change to a fraction.

$$8\frac{2}{3}$$

Change to a mixed number.

$$\frac{4}{3}$$

Change to a mixed number.

$$\frac{8}{3}$$

Change to a mixed number.

$$\frac{13}{4}$$

Change to a mixed number.

$$\frac{43}{7}$$

Make the fractions equivalent.

$$\frac{2}{3} = \frac{}{12}$$

Make the fractions equivalent.

$$\frac{3}{7} = \frac{}{14}$$

Make the fractions equivalent.

$$\frac{2}{3} = \frac{}{6}$$

Make the fractions equivalent.

$$\frac{1}{6} = \frac{}{6}$$

60,014	6 x 4	thousands	52,000
© CD-3756	© CD-3756	© CD-3756	© CD-3756
12,011	8,000 + 100 + 60 + 1	ones	1,300
© CD-3756		© CD-3756	© CD-3756
2,204	15,000 + 600 + 8	tens	4,090
© CD-3756	© CD-3756	© CD-3756	© CD-3756
3,340	3 x 6	hundreds	600
© CD-3756	© CD-3756	© CD-3756	© CD-3756

What is the numeral for sixty thousand, fourteen?

What is another way of writing (1 + 5) × (2 + 2)?

What is the place value of the 3 in 73,412?

What is 51,836 rounded to the nearest thousand?

What is the numeral for twelve thousand, eleven?

What is another way of writing 8,161?

What is the place value of the 8 in 58.36?

What is 1,256 rounded to the nearest hundred?

What is the numeral for two thousand, three hundred four?

What is another way of writing 15,608?

What is the place value of the 2 in 325.81?

What is 4,087 rounded to the nearest ten?

What is the numeral for three thousand, three hundred forty?

What is another way of writing 3 × (2 + 4)?

What is the place value of the 0 in 42,032?

What is 579 rounded to the nearest hundred?